Natural History

T0199827

The concept of 'natural heritage' has become increasingly significant with the threat of dwindling resources, environmental degradation and climatic change. As humanity's impact on the condition of life on earth has become more prominent, a discernible shift in the relationship between western society and the environment has taken place. This is reflective of wider historical processes which reveal a constantly changing association between humanity's definition and perception of what 'nature' constitutes or what can be defined as 'natural'. From the ornate collections of specimens which formed the basis of a distinct concept of 'nature' emerging during the Enlightenment, this definition and the wider relationship between humanity and natural history have reflected issues of identity, place and politics in the modern era.

This book examines this process and focuses on the ideas, values and agendas that have defined the representation and reception of the history of the natural world, including geology and palaeontology, within contemporary society, addressing how the heritage of natural history, whether through museums, parks, tourist sites or popular culture, is used to shape social, political, cultural and moral identities. It will be of interest to scholars and practitioners within heritage studies, public history, ecology, environmental studies and geography.

Ross J. Wilson is a Senior Lecturer in the Department of History and Politics, University of Chichester, UK. His research takes approaches from archaeology, anthropology, literature and sociology to examine aspects of modern history and its representation in the present. His current research examines the history and heritage of health and safety in Britain and the United States, the representation of the First World War in British culture, the history of New York, digital heritage, memory studies and the role of museums and heritage sites as a mode of social and political reform.

Natural History

Heritage, Place and Politics

Ross J. Wilson

Routledge
Taylor & Francis Group

LONDON AND NEW YORK

First published 2018 by Routledge
2 Park Square, Milton Park, Abingdon, Oxon OX14 4RN
605 Third Avenue, New York, NY 10017

First issued in paperback 2021

Routledge is an imprint of the Taylor & Francis Group, an informa business

Publisher's Note
The publisher has gone to great lengths to ensure the quality of
this reprint but points out that some imperfections in the
original copies may be apparent.

British Library Cataloguing in Publication Data
A catalogue record for this book is available from the British Library

Library of Congress Cataloging in Publication Data
Names: Wilson, Ross J., 1981- author.
Title: Natural history : heritage, place and politics / Ross J. Wilson.
Description: Abingdon, Oxon ; New York, NY : Routledge, 2017. | Includes
bibliographical references and index.
Identifiers: LCCN 2017005715 | ISBN 9781472470461 (hardback) |
ISBN 9781315597515 (ebook)
Subjects: LCSH: Natural history—Philosophy. | Nature—Effect of human
beings on. | Human ecology. | Nature conservation.Classification:
LCC QH14.3.W55 2017 | DDC 508—dc23
LC record available at https://lccn.loc.gov/2017005715

ISBN 13: 978-0-367-24412-5 (pbk)
ISBN 13: 978-1-4724-7046-1 (hbk)

Typeset in Times New Roman
by Keystroke, Neville Lodge, Tettenhall, Wolverhampton

Contents

List of figures

Acknowledgements

I would like to extend my deepest thanks to the individuals who have assisted in the completion of this book. Acknowledgements are due to the publishing team at Routledge, who have supported me throughout the writing process. I would like to express my thanks to the staff and students of the University of Chichester, Pace University and the New York Institute of Technology. My special thanks are due to those who have been influential in my work. It is the example of their fine scholarship and teaching which has served as an inspiration. Whilst any faults are entirely my own, any credit for this work should be shared with them. These individuals include Professor Julian Richards, Dr Dominic Perring, Dr Kate Giles, Dr Jonathan Finch, Dr Geoff Cubitt, Dr John Clay, Dr Kalliopi Fouseki, Dr Emma Waterton, Professor Laurajane Smith, Professor Richard Bessel and Dr Kevin Walsh. Finally, I would like to dedicate this work to my family. To my parents, Margaret and Roger, who always encouraged my studies, I will never be able to thank you enough. To Nancy, this book is for you.

1 Introduction

This work is a study of natural heritage and its representation in museums, parks, tourist sites and popular culture, to assess its definitions, its values and its uses within the contemporary western world. The central focus of the assessment is how sites of natural heritage in Europe and the United States have defined and shaped the ideology of the 'modern' age (Coates 1998). The representation and reception of this history has developed from the Renaissance, which saw the emergence of humanist ideals that placed nature as a subject of study rather than as a part of God's creation (see Morton 2007). This brought the study of the natural world to greater attention as evidenced by the formation of showcases, cabinets and botanical gardens of the gentry of the sixteenth century. These developments served as precursors to the collections of specimens which developed in Europe from the seventeenth century (see Oelschlaeger 1993). By the eighteenth and nineteenth centuries, the museum served as a means of exhibiting distinction as the arrangement of objects and displays evidenced both the advancements of the age and the order of animal society (Asma 2003). Sites of natural wonders also developed from the eighteenth century onwards as places where 'natural history' could be observed became locales of aesthetic, political and social reflection. This response can also be noted in the development of the conservation movement and the creation of national or regional 'parks' where 'nature' is preserved and protected for wider society. These specific sites, which were initially formed within the United States during the nineteenth century, offered another vision of natural history as a place which had to be saved from humanity's unceasing appetite for advancement. Therefore, from this perspective, the representation of natural heritage has been key in the formation of the modern world. Indeed, western society has defined and defended its status as 'modern' through its relationship with history of the environment (see Wilson 1991). This legacy has shaped the current status of natural heritage across contemporary western society. If it is through natural history that society has made itself then it is also through this heritage that individuals, groups and communities can be transformed for the future (after Harrison 2013; 2015). As the degradation of the natural environment gathers pace in contemporary society, increasingly it will be within the context of natural heritage that we ensure the continuation and enable the redefinition of humanity.

To undertake this work, this study applies an analytical perspective to the sites of natural history. Drawing upon the work conducted within the framework of critical heritage studies, it addresses how natural history, whether through museums, parks, gardens, tourist sites or popular culture, is used to shape social, political, cultural and moral identities (after Harrison 2013; Smith 2006). The study will assess how these sites, whether physical or 'media-ted', transforms individuals into witnesses who are called upon to testify to the significance of the object or event they observe (after Bird Rose 2004). The concept of the 'witness' is firmly embedded with Abrahamic cultures as a figure who is called upon to record the occurrence of an action. As such, the 'witness' is accorded a particular place within society as figure of importance, one who can provide insight, guidance and direction (see Hatley 2000). This is preserved in the legal system of western societies which regards witnesses with special distinction in the assessment of the burden of truth. This act of witnessing the natural world has been cultivated during the modern era as perceiving the detail and development of geological, climatological, botanical and faunal specimens has been a tool of difference and separation. To be a witness in this respect was to serve as a member of a select group where status was assured through this act of observance. However, the role of the witness is not one-dimensional, affirming the position, power and authority of the few. Whilst the witness can serve to support and affirm existing social and cultural structures, the witness can also possess a means by which hegemonic structures are undermined (after Haraway 1997).

The witness, in their role in recognising 'truth', can act as the centre of resistance and through their testimony give voice to alternative perspectives and agendas (Haraway 1997: 15). As such, recent scholarship which stresses the manner in which ideas about 'heritage' are constructed and contested through hegemonic systems of discourse will serve as a guiding principle for this investigation (see Smith 2006). Therefore, the historical, contemporary and future experience of the witness to natural history will be explored within this work (Harrison 2015). Sites of natural heritage provide distinct locales where witnesses are formed; individuals can observe the extent, beauty, difference and detail of the environment and then testify as to its meaning. It is these acts of witnessing which have become essential in the definition of the modern world and in the relationship between the past, the present and the future.

The recent recognition within critical heritage studies of a multidisciplinary approach is of particular importance for this work (see West 2010). A feature of new scholarship within this area of research is the extent to which 'heritage' encompasses a range of methodologies and theories (see Winter 2013). Therefore, to address these issues of place, politics and identity, this book utilises approaches from museum studies, anthropology, sociology, philosophy and environmental studies to assess the tangible, intangible, linguistic and intellectual relationships that contemporary society develops with the past to form an examination of the use of 'natural heritage' (after Harrison 2013). Through the use of visual, textual and spatial analysis, this study will focus its analysis on the formation of identities and ideals through the representation and reception of natural history within contemporary society. In this manner, aspects of natural history, the study of the geological and biological past and present, will be assessed as a system of 'witness' perspectives that are employed by institutions and used by individuals to define ideas about society, politics and culture. In this manner, the 'natural heritage' in Europe and the United States forms a means by which societies construct a sense of self and represent themselves to others. Far from being a remote and alien past, this heritage is made present through exhibitions, displays and popular representations; increasingly, it also serves as a means to impart a set of social norms upon society to address the risks and responsibilities that we may face in the years to come. This examination of natural heritage will detail how the past is represented but it will be focused on how this shapes the direction of individuals, communities, nations and the wider planet (see Harvey and Perry 2015).

Heritage, rights, responsibilities and natural history

The human impact upon the natural world over the last few centuries has been so drastic that it has radically transformed the global environment. The effects of industrialisation and rapid population growth have diminished natural resources, reduced and in cases removed wildlife and has increased levels of harmful pollutants, particularly the 'greenhouse' gas Carbon Dioxide (Sawyer 1972). Whilst the extent of climate change has been fiercely debated, an academic consensus was apparent by the 1980s which was reflected in the successive international treaties organised by the United Nations to limit the changes within the natural world caused by human activity. The United Nations Framework Convention on Climate Change (UN 1992) was negotiated at the Earth Summit in Rio de Janeiro, Brazil, in 1992 which agreed to stabilise the amount of Carbon Dioxide in the Earth's atmosphere. Further discussions led to the signing of the Kyoto Protocol in 1997 (UN 1997) and the Paris Agreement of 2015 (UN 2015), which called for a reduction in greenhouse gases to prevent significant and potentially catastrophic climate change. The rapidly developing concern of how to manage the environment represents the acceleration of these issues for contemporary society. Political groups and activists urge individuals, communities and governments to take action immediately to prevent irreversible damage to the earth's ecosystem. Therefore, this necessity for action, rapidity and advancement on these issues appears to be current across natural history and current environmental issues. These are the values that are placed upon this heritage by communities.

With the current environmental issues affecting people in every single society and nation state, this particular heritage stands as a vital component in understanding the direction of humanity and how it relates itself to the wider world (see Howard and Papayannis 2007). Whilst the popular engagement with natural history forms a significant component in the formation of the modern world it remains a highly understudied field within contemporary heritage studies (see Dorfman 2011). Indeed, even though a number of prominent assessments on preservation, management, legislation and protection of the natural world have been provided by scholars, a critical examination of the use and value of this specific heritage has been noticeable by its absence (Goudie 1981; Melosi and Scarpino 2004). This perhaps stems somewhat from the equation of natural history with a sense of 'natural resource' which is assessed on the basis of management, access and usage (Hutt, Blanco and Varmer 1999). This particular absence of concentration is replicated across similar fields and cogent disciplines, as whilst the development of heritage, tourist and museum studies have addressed how individuals, groups and communities respond to human history the critical heritage of natural history remains relatively unexplored (Barthel-Bouchier 2016). By examining how natural history is represented through media, political and public discourse alongside the contemporary displays and exhibitions that detail the evolution of plants, animals and the formation of minerals and the planet's climate, a new agenda regarding concerns for the environment, sustainability and the future within cultural heritage will be developed (see Lowenthal and Olwig 2006). It is through all these media that we witness our natural heritage as a constantly altering set of values rather than a 'natural' environment (Harrison 2015).

The way in which heritage sites shape and inform identities regarding power, place, rights and responsibilities has been expertly examined over the past two decades as scholars have argued that exhibitions and displays can engage audiences with progressive and affirming ideals (Sandell 2016; Sandell and Nightingale 2012). These concerns have largely been focused on a concern for human rights both as a transnational issue as well as at a local level with communities encouraged to accept and acknowledge difference within society (Murphy 2016). Such objectives have resulted in alterations within museological practice and a commitment from heritage sites to develop more inclusive displays which do not alienate or exclude the diverse audiences which constitute modern western nation states. Indeed, museums, galleries or heritage centres that focus upon transforming society, by addressing issues of prejudice and intolerance, are frequently regarded as 'sites of conscience' (see Aspel 2015). From former places of repression and torture, locations of inequality and injustice, to spaces of hardship and suffering, institutions based at these spaces use these historical associations to seek to provide education and guidance on how 'we' as individuals and 'we' as a society should live (Kidd et al. 2014). These sites assert the role of heritage as an object to think about the present and future rather than an object of the past to reflect upon. However, despite the increasing role that natural history museums, heritage parks and popular culture have had in recent years in addressing social and political issues, the study of 'sites of conscience' has remained firmly embedded within a modern 'human' experience (Cato and Jones 1991; Davis 1996). Scholarship on 'sites of conscience' has not been extended to natural heritage perhaps because of the very different relationship that is formed between humanity and the environment. However, in the context of a world which is increasingly beset by problems of climate change, an alternative paradigm must be formed to come to terms with a seemingly rapidly changing planet.

One area that has provided a means to extend this application of 'conscience' within this particular area of heritage studies is the formation of an approach within environmental and ecological thought which posits the significance of 'non-human' participants within society (Morton 2007; 2010; 2013). Within this method, human agents cease to be the sole participants in the world but instead take part in an 'object-centred democracy' where other agents operate alongside human counter-parts (Latour 1993; 2005). The implications for a critical natural heritage studies of such an approach is that the decisions on preservation, access and display are more than processes determined by humans but objects and events formed by the relationships society holds with a host of other non-human agents. The designation 'non-human agency' does not confer identity and being onto inanimate objects but rather takes into account that non-human actors exist and exert an influence on the world. It requires humans to take greater regard towards the wider environment and to move beyond the assumption of a superior role. In essence, this approach also requires a reorientation of the social to encompass 'non-humans'. This does not remove power from human communities; indeed this approach strengthens those communities as it facilitates the emergence of an 'object-orientated democracy' (Latour 2005). This comprehension of society in an alternative way enlarges what is included within that 'social' sphere and it also provides greater points of collabo-ration for human communities as discussions become focused on shared concerns rather than particular objectives. It is on this basis that sites of natural history can be regarded as places of 'conscience'; not necessarily where visitors are instructed with fundamental ideals but where values and ideals are presented, debated and defined. Natural heritage serves as a place where identities are formed and where societies can engage with issues that have implications beyond their own immediate concern.

Ecology, environmentalism and epistemologies

To develop a critical natural heritage studies requires engagement with the well-developed areas of research of ecology and environmental studies (Scheiner and Willig 2011; Worster 1994). These approaches have examined the relationship between human societies and the natural world to develop frameworks of manage-ment and to raise awareness of pollution and resource depletion (Keller and Golley 2000). The roots of these areas of concern are within the emergence of the conservation movement within the late nineteenth century where fears regarding the advancement of civilisation upon 'nature' saw the development of preserved and protected areas (see Taylor 2016). Within the United States, the development of state and national parks was based on the premise that a distinction existed between the 'human' and 'natural' worlds (Wellock 2007). The former was regarded as the artifice whilst the latter was unsullied by the deleterious effects of humanity. Indeed, it is within the work of the American transcendentalists Ralph Waldo Emerson (1803–1882) and Henry David Thoreau (1817–1862) that this sense of nature as a separate entity is best expressed. Emerson (1836) accorded nature with a redeeming quality whilst Thoreau (1854) defined the natural as a sanctuary against the incursion of the modern world. The marked division between humanity and nature which emerges during this formative period for the conservation movement forms a structuring device in the succeeding works on environmentalism throughout the twentieth century (Crane 2013: 219). However, whilst this initial separation was based upon a religious sentiment which placed an understanding and experience of the divine through engagement with nature, by the late nineteenth century the same point of difference is made but it is legitimated not through an appeal to God but to scientific endeavour. Indeed, it is during the course of the last one hundred years that humanity and the natural world have been cast as binary opposites on the basis of classificatory schemata.

Sites of natural history and the representation of natural heritage have been key in this process of dislocation as the environment became an object of the naturalist's gaze. During the latter half of the nineteenth century, specimens of fossils, minerals, plants and animals within museums and private collections were interpreted as evidence of development, order and progress. As such, natural history served as the basis for demonstrating the indubitable and inevitable character of a positivist, western, patriarchal and imperial knowledge system (Bennett 1995). The environment, separated from human experience, was catalogued in order to present it as an objective fact which could then assert contemporary social values and norms. Environmental and ecological science developed on this assumption throughout the course of the twentieth century. Indeed, this approach expanded from the 1950s as threats from industrial pollution, depletion of resources and climate change appeared to further justify the division between humanity and nature in order to preserve, protect and manage (see Carson 1962). By the 1960s, the emergence of environmental movements across the western world emphasised the need for an alternative relationship with 'nature' while serving to reinforce this division rather than ameliorating it. In this political, social and cultural movement, the environment and the natural world was still presented as distinct from humanity. Organisations such as Friends of the Earth (founded in 1969) and Greenpeace (founded in 1971) situated themselves as guardians of 'nature' against the threatening incursions of humanity. However, this division found its most distinct expression within the 'Gaia Hypothesis' which emerged in the 1970s as a means of explaining the Earth as a complex system which regulates itself and thereby fosters conditions for life on the planet. Whilst such an approach forwarded a holistic understanding of the environment, where geological, climatological and organic processes were entwined, it still maintained a divide between human agents and the wider ecological system (Lovelock and Giffin 1969; Lovelock and Margulis 1974). Whether as religious inspiration, scientific classification or environmental ethics, it is the separation of nature as an unassailable object, beyond interpretation. However, whilst the 'natural' is presented within these studies as neutral and objective it nevertheless has been framed entirely by cultural values (see Næss 1989).

As a means of addressing this divide between 'culture' and 'nature' which has been at the forefront of the environmental and conservation movement, the idea of nature as a social and cultural construct needs to be made prominent (Cronon 1997). Instead of assuming the environment exists as a stable entity, it can instead be regarded as an interpellated subject, already defined by ideology in the moment of its recognition (after Althusser 1971: 126). It is in this definition that an alternative mode of engagement can be formed within our understanding of 'nature' which examines the values we identify within nature and what the environment represents within society (Latour 2004a). Abandoning the division between humanity and nature is not a proclamation of an ecological or holistic mode of assessment where life on Earth is part of a system, for this is to assert a particular vision of the environment at the outset. Rather it is to acknowledge that our environment is constructed by humans and non-humans as a process of meaning-making, of a dialogue that shifts and alters as we seek to clarify not the nature of things but what these things mean to us (after Guattari 1989). Therefore, the 'natural' can be a place where we discuss how we might be rather than derive essentialist notions of what we are.

It is in this respect that 'natural history' becomes part of this process as its representation has been so integral in defining the character of humanity during the modern age. From the seventeenth century onwards the study of this topic has been central to understanding the self and the state (Descartes 1998). 'Nature' was equated with irrefutable claims to knowledge; concepts that were true in all times and in all places (Locke 1836). Immanuel Kant (1724–1804), within his 'Critique of Pure Reason', first published in 1781, defined the act of observance and the appearance of nature as evidence of humanity's innate sense of order:

> We ourselves bring into the appearances that order and regularity that we call nature, and moreover we would not be able to find it there if we, or the nature of our mind, had not originally put it there . . . (Kant 1998: 241).

Such notions of nature and the natural provided for the distinction between humanity and the wider environment. The imposition of humankind's will on nature demonstrated a desire not to be confined by nature but to ensure its rational use (Bentham 1823). The British Utilitarian philosopher John Stuart Mill (1806–1873) affirmed this relationship as the 'natural' was regarded as significant only through its mobilisation by wider society (Mill 1874). As such, beyond the human definition and utilisation of the environment, 'nature' possessed no inherent value in itself:

> Whatsoever, in nature, gives indication of beneficent design proves this beneficence to be armed only with limited power; and the duty of man is to cooperate with the beneficent powers, not by imitating, but by perpetually striving to amend, the course of nature – and bringing that part of it over which we can exercise control more nearly into conformity with a high standard of justice and goodness (Mill 1874: 65).

This concept of defining nature through use was also explored within the work of Karl Marx (1818–1883). Indeed, communism demonstrated that the material relationship between humanity and the environment was a key component in the structuring of society, the operation of power and the basis of political change. In this manner, it is through the association and engagement with 'nature' that modern, western society has orientated itself and defined itself. It is within our understanding of the environment that individuals, groups and nation states have become 'modern' (after Latour 2004b). The definitions of this oft-times nebulous term are varied but loosely interpreted this sense of modernity can be regarded as emerging from the seventeenth century as a response to the changing nature of capitalism, industrialism and urbanism. In this era of development, which differed substantially in places and time, a common theme of progress and advancement is discerned. The 'modern' can be characterised as a historical era which asserts its constant process of acceleration and development (after Noys 2014). 'Nature' and natural history serve as symbols of this sense of movement as it is in relation to this past and present that the modern age orientates itself towards the future (after Giddens 1991). The environment is thereby regarded as separate and stable in contrast to the dynamism and change that is put forward as characterising wider human society. However, to find within that natural history a preordained set of assumptions removes the potential of this history and heritage to affect change. In the assertion of the inevitability of progress, what is achieved is the maintenance of the ideology of that progress. In order that an alternative means of representing nature and natural history can emerge, an assessment of the dynamic agency of the environment and its perception can be forwarded.

In Friedrich Engels's (1820–1895) unfinished work, 'Dialectics of Nature', a vision of natural history as a process of motion, dynamism and effect is understood in the same dialectical framework that Marx and Engels developed for understanding the history of humanity. For Engels, natural history was the result of paradox and transformation just as society was the result of contradiction and change. This mode of understanding was also extended to the scholarship of the natural sciences, where Engels saw the unfolding of successive theories to understand geological and biological development as indicative of the movement witnessed in the relations of production within capitalism or feudalism:

> Dialectics, so-called objective dialectics, prevails throughout nature, and so-called subjective dialectics (dialectical thought), is only the reflection of the motion through opposites which asserts itself everywhere in nature, and which by the continual conflict of the opposites and their final passage into one another, or into higher forms, determines the life of nature (Engels 1940: 211).

If natural history, its representation and its perception are part of a dialectical process then the distinction between nature and society, stability and advancement, acts as a false dichotomy. It is through the recognition of the environment as a construct, formed from the agency of humans and non-humans, and representative of a system of values and ideals that we hold, that the definitions of 'nature' which have defined the modern world can be addressed. Nature and 'natural history' have been understood within the context of a positivist epistemological framework. However, 'natural heritage' can serve as a means of identifying places and spaces where dialogue and engagement takes place (after Latour 1987). To remove 'nature' from its vaunted position as a distinct and separate arena where humanity is an incursion is not to devalue the environment. Indeed, it is to include the natural world within the wider conversations that take place. As such, sites of natural heritage, from museums, media representations and tourist parks, form arenas where new visions of the world can be witnessed.

The representation and reception of the natural heritage and history across western society provides a means of demonstrating and performing social, cultural, political and moral identities. It also enables individuals, groups and communities to form a distinct 'sense of place' within their locality, from a national context but also across a wider, transnational agenda (Schofield and Szymanski 2011). Natural history acts as a means to express ideas and values beyond the confines of a time-specific era; this particular heritage can shape current society by placing it into a wider perspective (Heise 2008). From the formation of the earth, the creation of the continents, the emergence and extinction of the dinosaurs to the ice ages that sculptured the landscape and wildlife that constitutes our current environment, the natural history of the planet constitutes a unique type of heritage. To engage society with this past provides a means of educating and inculcating a set of behaviours that can alter the definition and perception of humanity. This study does not assume that notions of 'natural history' exist solely within the museum or heritage sites where visitors are informed of their place within a vast timescale of global development. Rather, it assesses how this 'natural heritage' constitutes part of a wider popular culture as aspects of an ancient past, whether geological, climatological or palaeontological, are incorporated within theme parks, language and media representations. This assessment of the place and value of natural history in contemporary western society identifies how this heritage is used by institutions, corporations and the wider public to shape and inform the present.

Witnessing natural heritage

To examine how sites of natural heritage function within society to create spaces of engagement, the way in which these places provide a means of witnessing will be the focus of this assessment. The perspective of the witness will be used as it provides a point of reflection on the way we place values onto objects and scenes. To be a witness is an active commitment with and orientation towards the wider world, to observe, to reflect and to testify as to the occurrence of what has been seen, heard or experienced (after Dewsbury 2003). The figure of the 'witness' is present within a variety of contexts across western society and it is frequently regarded as the bearer of truth. Within a legal context and with a religious connotation, the witness is one to whom the 'truth' has been revealed and it is this which establishes the privileged position of the witness and the duty to testify. However, the etymology of the word equates the term to personal knowledge rather than unassailable truth, derived as it is from the Old English 'witt'. Rather than the 'witness' possessing a singular truth that they are obliged to render up to their wider community, the role of the witness indicates the significance of active participation within the world (Ricoeur 2004: 264–265). The witness is also a term that does not necessarily equate with a human participant. We can speak of places, spaces, objects and animals all bearing witness to an event or occurrence. Therefore, to be a witness is to participate in the 'democracy of things' as conceived by Latour (2005). The 'witness' in this context is not an objective recorder but a figure whose perception is acknowledged to be specific in time, individual in scope and singular in experience (after Lyotard 1988: 26–27). What is significant about the witness is their role towards the event; the manner in which their attitudes, ideas, values and identities are formed in relation to the action with which they have engaged (Thrift 2000). To study the 'witness perspective' requires an analysis of the attribution and formation of meaning not passive affirmation (Thrift 2003). To act as a witness places moral, social and political obligations onto the individual through forming a connection between the individual and the event itself, as they are required to bear the burden of their witnessing and to testify to its significance (after Haraway 1997).

To witness natural history within the museums, parks, heritage sites and wider media is to be orientated within a wider context of other human and non-human agents and to assert an understanding of that observance. These acts of witnessing take place within a variety of contexts but they are always conducted for a purpose. This emphasis on agency and activity is significant as the witness perspective enables the assertion of an alternative approach to how heritage is conceived and used within society. From the emergence of studies which were critical of the 'heritage industry' within the 1980s, there has been a distinct concern that the past is 'sold' by institutions and 'consumed' by the public (Hewison 1987; Lowenthal 1985; Wright 1985). Heritage has been defined and understood as a concept purely within this consumerist model of study which has structured how the relationship between individuals, groups and societies and their sense of the past has been assessed (after Harrison 2013). This particular framework of analysis does assume a passive role of the 'consumer' within its assessment as notions of agency and engagement are peripheral whilst a vapid consumption of the 'heritage product' takes place (Smith 2006). The hegemonic process of heritage production is clear but the model of consumerism does not adequately define how the past is mobilised, experienced and understood (after Harrison 2013). Therefore, in recent years a number of scholars have employed alternative schemes which assess the emotion, affect, agency and engagement within the use of heritage (Smith and Campbell 2015). This has succeeded in offering alternative models of analysis beyond the consumer thesis. The application of the witness perspective within heritage studies furthers this agenda as it provides an approach to heritage which is not reliant upon a commercial structure as a pre-conceived means of assessment. Instead, this notion of witnessing demonstrates the active engagement that exists within the relationship to the past and its role in defining the future (after Harrison et al. 2008).

This orientation towards what still is the 'yet-to-come' is also significant in the role of the witness. One of the features of the advent of the modern era has been the sense of imposing order and understanding upon the imminent. As the present is structured by a perception of advancement and development, this has altered how the future is regarded; as an unknown variable, the 'what-will-be' is cast as an inevitable extension of the current era. Different political, cultural or social structures are not countenanced as it would irrevocably disrupt the teleological progress that forms the basis of late western capitalist culture. In essence, this is a process of 'colonising the future' as the potentially different and dissonant imminent is rendered into a safe and affirming objective (Giddens 1991). Heritage becomes a significant part of this process with the representations of the past acting as evidence of this progress and development. Museums, archives and libraries through displays, exhibitions and arrangements appear to demonstrate the movement towards an ever-accelerating present (after Noys 2014). Indeed, the edifice of the modern age has been formed through this use of heritage and this relationship towards the past (see Bennett 1995). To counter this consoling and conservative aspect of heritage studies, a number of scholars have asserted the dissident voices present within the past that can destabilise contemporary society and its sense of advancement (Smith 2006). Similarly, other accounts have stressed the need for heritage studies to be more 'future orientated' to ensure that the linear narratives of progress are disruptive as a sense of the past is used to alter what might be (Harrison 2013). It is the framework provided by the witness perspective that can develop these approaches as the witness offers no guideline or direction for development. Indeed, the witness speaks of their observance as a means of ensuring that events are not ignored or made liminal by a movement to progress (Bird Rose 2004). The witness can anchor an event or action in space and time and as such alter and reform the past, present and the future.

This study of natural heritage utilises these approaches to demonstrate how representations of natural history within contemporary western society form a means of witnessing. Within the spaces, places and sites of natural heritage across the United States, Canada and Europe, individuals, groups and communities are required to bear witness to objects, events and processes that alter personal and collective identities and can amend or reimagine the way in which societies could be organised. These representations form a range of witness perspectives that each provide an alternative view on the world. This occurs within the sites formed by museums, parks, heritage sites and the wider media where this engagement with natural heritage takes place:

> Places are witnesses, locales of memory that we mark out or that simply are there waiting, traces that serve to remind us of those things that need remember, for which there is a duty of one kind or another for us to bear witness (Booth 2006: 111).

These places enable the act of witnessing to occur where those present can individually or collectively testify to the events or actions that they observe. The role of the witness within these spaces can be generalised within three separate categories that provide a framework to assess how sites that represent natural heritage engage wider society:

> Moral witness: in this perspective the representation of natural history is structured to impart a moral lesson. Whether through the arrangement or display of objects, the presentation of data or the long-term time perspective provided by natural history, the affirmation of a moral identity is made within these sites.
>
> Political witness: this mode of witnessing natural heritage constructs the engagement with this past and present around issues of local, national and international political concern. Whether focused on issues of democracy, sustainability, capitalism, socialism, environmentalism, power and equality, natural heritage sites are used to form political identities.
>
> Social witness: within this type of witnessing natural heritage represent-ations, sites form a means of effecting social identities in terms of gender and class but also affirming national or local identities. Through the structure and presentation of this heritage, notions of social norms and behaviour can be asserted or altered and potentially new forms of society can be considered.

Through these categories of witnessing, the representation of natural heritage across contemporary western society can be examined as an engagement with time, place and identity (after Deleuze 1988). Heritage sites create places where the witness can observe the processes of natural history and testify to its relevance. It is this act of witnessing that ensures the significance and role of this past in cur-rent society. Places of natural heritage create witnesses, which shape the present and guide the future through the representation of the past. It is this relationship between the historical, the contemporary and the imminent that marks natural history as key in understanding the modern world. As these witness perspectives are demonstrated within sites of natural heritage, the tensions, contradictions and aspirations of society can be observed. It is within the relationship to the environmental past that western society has defined and realised itself. The repre-sentation of natural heritage has been key in shaping social relations, cultural identities and political ideas within the modern era. In the context of current con-cerns regarding the environment and sustainability, it is through the engagement with natural heritage that society will orientate itself towards the future. As such, the acts of witnessing made at sites such as museums, heritage parks or television programmes, will provide a context for how individuals, groups and communities regard their place within the world. The function of the witness is to observe the past, testify to the present and to assess the future. However, the relationship that is formed with 'what may occur' is not the act of colonisation of the future which is regarded as characterising modernity (Giddens 1991). The witness testi-fies to their understanding of the world, a testimony that exists in connection and communion with the wider environment. It is within this dialogue that alter-natives emerge and the inevitable appearance that the imminent may possess can be disputed (Latour 2004a). Therefore, this study will provide an assessment of how natural history is represented and its effects on the understanding of the past, present and the future. This is undertaken as a means of an active, critical engagement with this particular heritage which, whilst appearing stable, has shifted in form and meaning to support notions of power, place, authority and identity.

Chapter outline

This analysis will be conducted over four chapters; each of these details the ways in which the representations of natural history have constituted modern society. It will assess the manner in which these sites provide modes of witnessing which impact upon how the world is perceived. The history of these representations will be assessed at the outset before the sites of engagement with natural history, in terms of museums, national parks, heritage sites and media representations, are examined. The legacy of how this heritage has been represented is as much part of this process as the act of representation in the present. These witness perspectives do alter in time and context but it is in the act of witnessing that alternative or affirming perspectives are formed (after Haraway 1997).

Chapter Two provides a critical assessment of the representation of natural history from the sixteenth century to the present day. Within this chapter the establishment of natural history as part of western cultural identity will be assessed through a study of museums, parks and gardens, detailing how the study of the natural past has been developed during the modern era. In this initial examination, the significance of natural history in defining the modern world will be emphasised. Ideas of capitalism, nationalism, ethnicity and gender have been formed through the representations of natural heritage. Therefore, this chapter will examine the earliest accounts of representing natural history which emerged within the trend for collections and displays of noblemen during the sixteenth century. The specimen and curiosity cabinets across Europe during this era were used to establish order and reason upon a chaotic world, to marvel at divine creation but to place man at the centre of this scheme. As such, natural history was used as a mirror for the social values of the age. This study will then track the development of this process to the formation of institutions and organisations which flourished in the seventeenth century, as the examination and classification of the natural world was perceived as a means of social and intellectual progression. The intellectual status and conspicuous interest amongst elite social circles for the study of natural history formed the basis of eighteenth century involvement within natural heritage. During this era the geological and fossil record was frequently evoked within Europe as a metaphor or allusion for political development as 'revolution' or 'stasis' was defined through the appearance and study of the environment.

By the nineteenth century, across Europe and the United States, the formation of natural history museums, societies and journals served to form or affirm national identity. Through a connection to the ancient fauna and flora of a particular place and its development to the present day, a formative narrative was created for nation states seeking to assert a political and cultural identity. This era also saw the development of exclusory accounts of natural history that classified issues of development and progress and used this to understand wider human society. Darwin's evolutionary theory was mobilised to this effect as institutions used this scheme to organise natural heritage in a manner that supported racial prejudice and segregation. A focus in this chapter will be upon the expositions and museums that were formed during the latter half of the nineteenth century which reflected this constructed vision of natural heritage and catered to the growing public interest in the field. The formation of what constitutes a 'natural history' during the nineteenth century reveals the manner in which this apparently stable part of the development of the world is in reality the product of social discourses and political perspectives. The focus on development, progression and the association with the remains of ancient plants, animals and rock formations as principles of nationalism, gender, class and behaviour demonstrates the ongoing involvement of natural history in forming identities in the present. This chapter will, therefore, introduce the key element of this study within the analysis of this chapter: that natural heritage is dynamic. This aspect of our contemporary world that may appear so permanent has been physically, figuratively and literally changing as modern society responds to the anxieties and risks of their current age. Through this chapter, the representation of natural heritage in the present can be observed to be a palimpsest of previous eras which will itself pass onto another mode of witnessing as we seek to draw new conclusions from what is at times ancient contexts. Natural heritage has formed the modern world, its ideals and its identities. To be a witness to this process is to observe the enactment of modernity. In an era of change and uncertainty, it is this relationship with natural history which will enable alternative perspectives on the past, present and future to emerge.

Chapter Three will focus on the representation of natural history within contemporary museums and educational sites within western society to demonstrate the construction of witness perspectives at these places. Using the three modes of witnessing – moral, political and social – the engagement of society at these sites will be assessed. Examples from the United States, Britain, France and Germany alongside other nations in Europe will be used to assess how the objects, categories and timeframes present within exhibitions and displays serve to frame the relationship with natural history. Therefore, this study will detail how current museological practice structures and uses this heritage within society. This approach is significant as it further demonstrates the significance of the witness as individuals, groups and communities are presented with a means of perceiving the development and direction of the environment. In this manner, to further demonstrate the malleability of this heritage, the way in which aspects of natural history are presented to form national, regional, political and moral identities will be the focus of this analysis. As such, the chapter will provide a valuable catalogue of national and provincial institutions from across contemporary Europe and the United States that represent the geological, climatological and palaeontological heritage of places, territories and nations.

This work will be contrasted with the array of regional heritage sites in towns and cities that utilise natural history to both tell the story of a place and to shape the body politic. Local sites in Europe and North America will provide further examples of the way in which natural heritage is structured for particular effect. These places provide points of connection to ancient and modern history that informs but also exercises and extends the perspective and judgement of those who witness these exhibitions and displays. In this way, the representation of natural history will be examined as to the way in which it structures the witness perspective and informs the identities of their visitors. Whether national, regional, political or moral, the display of ancient to modern animal and plant life is used to frame current concerns and agendas. The arrangement of exhibitions, the display of objects and the narratives used all provide a means by which visitors are provided with particular witness perspectives on the environment in the past and in the present.

Chapter Four will examine sites of natural history and the formation of a sense of time and place at these locales. Across Europe and North America, natural history is represented within open-air sites that display geological anomalies, ancient landscapes or places of prehistoric animal habitation *in situ*. From the inspiring vistas of the Grand Canyon in Arizona, United States, to local natural history trails in the Scottish Borders, these places are a means of connecting past, present and future through natural heritage. Through the use of the witness perspective, the manner in which these sites engage visitors and orientate movement around spaces of natural history will be examined. These places demonstrate the significance of natural history for contemporary concerns as they can serve to render human experiential timescales insignificant whilst enabling the formation of intimate attachments between individuals and communities with specific places. From the sublime sights of extensive ancient rock formations to the humble identification of sedimentary layers and geological principles, this is a heritage that acts upon its visitors with significant consequences. Therefore, whilst these places may appear seemingly disparate in scale and although their presentation may alter from region to region, nation to nation, these places of natural heritage structure notions of place, power and authority within contemporary western society. The context of this natural history is significant as it forms a specific relationship with the visitor as a witness to such experiences. These places can be easily disregarded within assessments of 'natural history' as they appear 'untouched' by humans. The vaunted position of Yosemite National Park, California, United States, or Exmoor National Park, Somerset, England, as preserving natural habitats are two examples of this notion of an 'untroubled' wilderness. The apparently 'neutral' perspective that can emerge at such sites that are without extensive development or guides does not entail the absence of power and representation. Indeed, such places concentrate the relationship between wider society and notions of 'natural heritage'. It is within these sites that values and ideas are exposed as visitors are asked to witness the significance of this heritage for themselves. This serves to focus ideas about the meaning of 'natural history' in the present and for the future.

These *in-situ* sites engender a connection to place, region, values, morality and politics within contemporary nation states by enabling visitors to bear witness to the forces that have shaped physical landscapes but also cultural identities. Natural heritage within these locales inspires and defines individuals, communities and wider societies as it places them within wider timescales, alternative eras and different visions of the present. As part of this broad study, sites and landscapes will be a core part of this analysis. The natural environment does not constitute some passive backdrop to cultural heritage sites, rather it constitutes a highly valuable aspect of cultural heritage. The significance of these places will be explored within Europe and North America on a local and national scale, utilising the legislation, exhibitions, information boards, guided pathways and interactive elements within these places but also the trails, tours and setting of locales of natural heritage. For example, well-developed visitor sites such as the Giant's Causeway in County Antrim, Northern Ireland, will be examined alongside the display and management of the Jurassic Coast across Dorset and East Devon in southern England. Similarly, isolated sites such as Siccar Point in Berwickshire in the Scottish Borders can be contrasted with Hookney Tor in Dartmoor as places where visitors witness the ancient formation processes that shaped the British Isles without significant interpretative intrusion. The manner in which this particular heritage is communicated informs society about their own situation in the present; as guardians of the environment, as part of a wider ecological system, as mere points within the wider history of the planet's development. The representation of the past within these sites of heritage is key in forming significant notions of identity within contemporary society. To witness natural heritage within these locales is to engage in a dialogue about the current issues and future concerns.

Chapter Five will extend the study of natural history by assessing the role of tourism and popular culture within this area of heritage. Whilst eco-tourism has become a well-established field of enquiry within tourism studies and geography, the role of tourism within the study of natural heritage has not been fully established. Therefore, this chapter will provide an assessment of how visiting tourist sites representing natural history structures modes of witnessing. Tourist sites are distinguished here from the museums and *in-situ* places of natural history as locales where this heritage is mobilised for the purposes of commercialisation. This is not to suggest that museums, heritage sites or national parks do not have commercial interests or serve the interests of tourists; it is to note the function of specific places as orientated entirely toward popularisation and profitability. Whilst some of these tourist sites are state-sponsored, many are private enterprises developed solely for capitalist purposes. For the purposes of this study, 'dinosaur parks' will form the basis of analysis within the first section of this chapter. Whilst providing an economic benefit to communities and societies, these tourism sites also provide specific points of reference to witness the place of natural heritage in the present. These locales detail for wider society the significance of natural history through displays, recreations and multi-media installations. Whether it is the retelling of the story of the evolution of the dinosaurs or detailing the habitats and lifestyles of these animals across the millennia, these sites connect tourists with a distant past and make it part of current concepts of heritage. This discussion will analyse the content and form of tourism sites which engage visitors with histories of the fauna and flora of Earth.

In these settings, natural history is firmly commercialised but it is also politicised. The way in which these tourist sites use the past is distinctive – from creating active, engaged citizens concerned about the environment through to the formation of 'consumers' of this heritage, actively engaging in reducing this past to a series of identifiable ciphers which reaffirm notions of identity, place and power within society. Within these spaces of tourism in Europe and North America, witnesses to this heritage are formed which have a significant impact upon how individuals and communities relate to the environment in the past and the present. It is this response to natural history which will be explored within this chapter as the place and value of this heritage within popular culture is assessed. Alongside this assessment, the wider appearance of natural heritage within popular culture will also be assessed. Therefore, from company logos which draw upon this heritage as a sign of strength and adaptability, advertising campaigns using extinct prehistoric mammals to promote the effectiveness of their products, to the myriad of ways in which natural history forms part of everyday life within political, media and public discourse, this study will reveal the tangible and intangible heritage of natural history. Rather than drawing complex distinctions between commercial and popular understandings of natural history, the study will focus instead on what this heritage is used for. Moving beyond a simplistic model of consumption, this chapter will argue that natural history is used to provide perspectives on society (after Guattari 2008). This heritage is mobilised to evoke a sense of power, politics and place which demonstrates that, far from forming a neutral part of the past, natural history is at the forefront of current debates surrounding identity. The use of this heritage enables the formation of witnesses to how natural history structures contemporary society.

The book concludes by reiterating the importance of a critical heritage agenda in the examination of natural history in museums, heritage sites and wider popular culture. This approach provides an alternative view on the use and value of natural heritage, not as a fixed resource that requires preservation and protection or a permanent backdrop to human activity. A critical approach to natural heritage reveals the malleability of this aspect of society, culture and politics. Rather than considering this solely as a point of critique, this can enable a recognition of the act of witnessing at these sites which can build a democratic space between the environment and human society. Crucially, it is within this space of witnessing that identity, politics and power are negotiated not with the natural history as a passive element but as an agent in the process of representation. Witnessing provides a framework of assessment that does not place human interpretation as the sole point of interpretation. This mode of analysis enables the recognition of a changing relationship with nature and the natural world throughout the past (Guattari 2008). The creation of this interpretative arena where witnessing enables recognition of natural heritage and the relationships that are formed with the environment by human societies can assist in developing solutions to current issues. As a changing climate and environmental degradation alters the way in which groups and societies interact with one another and the wider world, viewing natural heritage as a product of changing relationships not static essentials can ensure the adaptation and cohabitation of human society and alternative visions of 'nature' (Morton 2016). Natural heritage is key in this process as it provides a means of understanding the past, defining the present and, hopefully, ensuring a future.

2   The representation of natural history

Introduction

At the outset of 2015, the Natural History Museum in London announced that it was to remove the imitation dinosaur skeleton of a Diplodocus (*D. carnegii*) from its Central Hall where the twenty-five metre model dominated the vision of those entering the institution (Figure 2.1). This cast copy was made from a specimen unearthed in Wyoming in 1898–1899 which was bought by the industrialist Andrew Carnegie (1835–1919) for his new institution, the Carnegie Museum of Natural History, which had opened in Pittsburgh in 1896. Whilst other species had been discovered in the preceding decades, which had witnessed an enthusiastic pursuit of fossils, this example was regarded as a distinct class within the family *Diplodocidae* and named after its wealthy owner. The initial excavation of the fossilised remains was widely reported in the press, which marvelled at the sheer size of the creature. Such was the renown that Carnegie had reproductions made of the original for King

Figure 2.1 Diplodocus carnegii. A detail of the replica Diplodocus carnegii skeleton situated in the central hall of the Natural History Museum, London © The Trustees of the Natural History Museum, London.

Edward VII (1841–1910), who donated the model to the British Museum (Natural History) in South Kensington, London. Further replicas were made for museums at the requests of monarchs and heads of state in Berlin, Bologna, La Plata, Madrid, Paris, St. Petersburg and Vienna. The London Diplodocus was unveiled in 1905 in the Reptile Gallery to crowds of visitors eager to see this ancient animal. After the formal separation from the British Museum (Natural History) to the Natural History Museum in 1963, the model was repositioned with a horizontal neck and a whipped tail to reflect the revolutions in palaeontology during this period (see Bakker 1975). Studies had reclassified dinosaurs as warm-blooded, social and active creatures rather than the cold, reptilian, lumbering beasts of popular imagination. With the increase in public and scientific fascination in dinosaurs during the 1970s, the model was relocated from its original display to stand in the Central Hall in 1979. In 2014, the £5m donation from Sir Michael and Lady Hintze to the museum was reflected in the renaming of the grand exhibition space as the Hintze Hall and plans were set in place to remove the model and replace it with the skeleton of a Blue Whale which had itself been in the institution's possession since the late nineteenth century and displayed since the 1930s. The public response to the news of this alteration was immediate as outrage was exercised within the headlines and columns of the nation's newspapers (Anon 2015; Singh 2015).

The model, which had acquired the nickname 'Dippy', a moniker it shared with the original set of fossilised bones in Carnegie's museum in Pittsburgh, had appeared to have gathered such a cherished place within the affections of British society over the thirty years in which it had been centrally displayed that it seemed a permanent fixture. To remove and to relocate what had become such an established part of national life appeared to be tantamount to the destruction of a vital part of Britain's cultural heritage. Sir Michael Dixon, the Director of the Natural History Museum, attempted to ameliorate the sense of loss and disappointment by asserting the status of 'Dippy' as a model which, although albeit an apparently beloved model, did not provide modern visitors with an accurate view of the institution's role. This was contrasted with the effect that the skeleton of the Blue Whale is supposed to induce. The plan is to have the animal's bones displayed in the hall to the amazement of the public as if it was about to embark on a dive, with jaws ajar, ready to feed. Such a dramatic scene has been chosen to communicate the relevance of the Natural History Museum within contemporary society and its function in studying and conserving the natural world. It would appear that the model diplodocus, once the cutting-edge of scientific research, had been regarded as extinct with regard to understanding the environmental pressures faced across the globe. The shifting values we place upon natural history are reflected in this account. The reproduction diplodocus fossil has been the subject of an industrialist's pride, the object of scientific study and debate, an emblem of internationalism, a means of education, a mode of entertainment and an item of sentimentalism. This constant process of redefinition stands in stark contrast to the appearance of permanence that this display conveys. Indeed, the response to the removal of the dinosaur was born out of an assumption of its presumed longevity as an exhibit and the familiarity with which it was held. The conjecture that the looming presence of the Blue Whale, a creature still remote from everyday human experience, will be able to better convey the institutional message of the museum is also formed from notions of immediacy and effect. As such, we are presented with the museological display of natural history, not as a permanent bedrock of our heritage, but as an ever-shifting system of values and ideals. Our natural history is a social construct that reflects our desires, our anxieties and our identities in the present and forward into the future.

The way in which natural history is subjected to the fluctuating notions of culture and society has been well-attested by historians who have examined the emergence of an interest in the field from the seventeenth century. The study of natural history was defined from the 1600s by the gentlemen scholars whose collections of natural wonders served to affirm social distinction but also a humanist perspective upon the world. In this manner, classifying natural history provided evidence of humanity's capacity of rational agency beyond the dictates of a divine power (Oglivie 2006). The development of the study of natural history during this era forwarded an image of this past as unassailable and immovable. Therefore, its assessment by early scholars firmly embedded values of the nation state upon items or places of natural heritage. To observe this heritage was to recognise the longevity of the nation itself. A sense of permanence was attributed to the environment as a testament to the legitimacy of government. This character of natural history was further established within eighteenth century collections, where specimens of flora and fauna from across the globe were used to normalise colonial rule. Displays and exhibitions of this heritage provided a means of placing control and order upon a 'disordered' world. European collectors were able to utilise the character of natural history as a means of exploring and exploiting the resources of colonised regions. Natural history is redefined in this manner to enable the wider ambitions and to reflect the wider anxieties of society.

This is demonstrated in the growth of tourism to 'wild' sites in the eighteenth century, where visitors could see the spectacular character of natural history. From deep gorges, waterfalls, rock formations and mountain views, a 'sublime' character to this heritage was ascribed to these places as in the maelstrom of the industrial revolution the natural world offered a place of reflection. By the nineteenth century, natural history had maintained its status as a subject of taste and distinction as studies of the development of animals and plants were the pursuit of the educated and wealthy. Upon the features of the environment were placed social and cultural values that elevated it as a topic of respectability. In the twentieth century, the representation of natural history has been used to conduct wars, define gender roles, negotiate issues of ethnicity and reassess the notions of identity within contemporary life. In the twenty-first century, natural history stands as a means by which wider society can address the pressing issues of environmental change and rethink the relationship between the human and the non-human. Therefore, to assess the status of natural heritage, its uses and its values, a recognition of its constantly shifting status needs to be made.

The origins of natural history: classification and distinction

The interest in the natural world emerged during the Renaissance as Humanist ideals brought greater regard for an environment measured by 'man' (Imperato 1599) (Figure 2.2). Scholars would accumulate cabinets of fossils, specimens and curios as a means of studying the intricacies of nature and as a personal reflection of status and learning (Tradescant 1656). Such collections were at the forefront of colonial endeavours, as new plants and animals attracted a sense of wonder and excitement (see Rumphius 1705). As these assortments grew in size and scope, they became housed in larger display halls that were open to sections of the public for viewing (Grew 1681). In Britain, exhibitions of natural curiosities from the early eighteenth century onwards drew large crowds eager to examine the assemblage of fossils, shells and skeletons that constituted such collections (Arnold 2006). For example, the Leverian Museum opened in London in 1775 as an attraction for polite society, providing a venue to amuse and to educate (Anon 1790). Drawn from the items gathered by Sir Ashton Lever (1729–1788), this commercial venture was formed to fascinate the visitor. Similarly, the Museum Richterianum in Leipzig opened in the 1740s to show the collections of the merchant Johann Christoph Richter (1689–1751) (Hebenstreit and Christ 1743). Indeed, in Germany and France, collections of specimens had been publicly exhibited since the beginning of the 1700s with societies forming in capital cities and across the regions to

Figure 2.2 Display room of natural history collection. Ferrante Imperato (1599), Dell'Historia Naturale. Naples: s.n.

promote the study of the natural past (Jardine, Secord and Spary 1996; Spary 2000). The Gesellschaft Naturforschender Freunde zu Berlin (1775) was founded in the 1770s to disseminate the works of scholars for wider appreciation, whilst the Société d'Histoire Naturelle de Paris was founded in 1790 to encourage research in the field of geology and palaeontology (Pinel 1791). It was the work of groups such as these that popularised a discourse of science and aesthetics regarding natural history during the eighteenth century (Vuillemin 2009).

In this manner, the ordering of flora and fauna into taxonomies created an object of admiration in itself; to form the diversity of the natural world into a comprehensible system, to render it knowable, was a form of appreciation. This was encouraged by the classification of the natural world through the defining work of the Swedish botanist Carl Linnaeus (1707–1778). The naming system of classes, orders, families, genera and species which ranked organisms by attributes forwarded by Linnaeus (1735) in *Systema naturae* also conceptualised the environment as an ordered entity that science could place within a framework. Linnaeus had developed a museum collection to emphasise the relationships within this order which was exhibited from the 1860s (Linnaeus 1764). The influence of the Linnean system was substantial with societies of gentleman scholars forming across Europe to study natural history in a systemic manner from the 1780s onwards. An ordered natural history, based on class and hierarchy, appeared to affirm the social ordering of nation states whilst disruptions to this order and expressions of nature's upheaval could inspire dissent and revolution (Miller 2011).

The consequence of this use of geology and palaeontology was the emergence of the taxonomic or classificatory museum which dominated natural history displays from the beginning of the 1800s. The nineteenth century natural history museum operated as a canvas upon which to project a vision of ordered society (see Flower 1898). In Australia, Britain, Canada, France, Germany and the United States, local and national museums of natural history were formed during this era as a product of social relations and the perception of the values of this past (Yanni 1999). The 'national' museums of natural history were founded during this period. In Berlin, the basis of the Museum für Naturkunde was developed in the early 1800s at the Humboldt University of Berlin and formalised by the mid-nineteenth century. In Vienna, the Naturhistorisches Museum was formed out of late eighteenth century collections and established as a separate entity in 1876. This followed the orders of Emperor Franz Joseph I (1830–1916), who envisioned the museum as a demonstration of the Hapsburg Empire's power and status: the order of one kingdom is reinforced through the study of another. Indeed, the museum bears the engraving: 'Dem Reiche der Natur und seiner Erforschung' (The realm of nature and its exploration).

The 'kingdom' of nature is an essential part of this use of environmental history and can be observed in the earlier formation of the Kunstkamera in St. Petersburg in 1727. This institution, which was reformed during the latter part of the eighteenth century, enabled the display of natural curiosities as a means of validating the rational function of the state within Russia. The relationship between natural history and the body politic is most clearly notable within post-revolutionary

France. Whilst royal collections of specimens had been formed in the eighteenth century, with a grand menagerie at the Palace of Versailles the centrepiece of the Bourbon demonstration of order in both the state and in nature, this was overturned with the revolution of 1789. The objects of the collection were not disposed of as the trappings of the decadent regime. Indeed, they were reimagined as the ownership of the people and evidence of the principles of revolution:

> La ménagerie de Versailles qui, sous l'ancien gouvernement, ne fut pour la nation qu'un objet inutile de luxe et de dépense, a été transférée au Muséum d'Histoire naturelle pour servir à l'instruction publique (The menagerie of Versailles which, under the old government, was for the nation only a useless object of luxury and expense, was transferred to the Museum of Natural History for use in public instruction) (Toscan 1795: 6).

Accordingly, the Muséum national d'histoire naturelle was established in 1793 and served as the centre for research into ancient flora and fauna. Within the displays and the work of the pioneering naturalists who were employed in the institution, natural history was represented as a means of understanding the new social and political order within France. Georges Cuvier (1769–1832) was appointed to the position in the museum by 1798 and proceeded to formulate his hypothesis of catastrophisme, where new life adapts and improves in the wake of grand events (Cuvier 1813). It was Cuvier's associate in the museum, primarily Jean-Baptiste Lamarck (1744–1829), whose focus on adaptation altered the perspective on natural history (Lamarck 1801). For Lamarck, organisms were the product of inherited adaptation, the result of advantageous changes made by progenitors that are bequeathed upon descendants (Lamarck 1809). In an era of political turbulence, with the Republic founded in 1793 being succeeded by the First Empire under the Consulship of Napoleon Bonaparte (1769–1821) in 1804, Lamarck's evolutionary model provided assurance for this upheaval. The narrative of natural history within this work and the museum mirrored the alterations within early nineteenth century France (Burkhardt, Jr. 1977).

The classificatory museum was used in this manner to portray the movement of natural history as a means of understanding the fluidity of society after 1800. This process was witnessed in the regional and national museums within Germany during the nineteenth century as rapid industrialisation and rising nationalism were addressed within displays of natural history (Nyhart 2009). Museums were founded in Hamburg, Stuttgart and Frankfurt during this period which all exhibited collections that reinforced notions of heimat through identifying the origins of flora and fauna in the Vaterland. In Hamburg, the scientific society, Naturwissenschaftlicher Verein, was behind the forming of the Naturhistorische Museum in 1837 (Anon 1846). The collection within this institution reflected the industrial development taking place within the city as the port city expanded; just as the animal kingdom adapted and altered so too did the economy of Hamburg (Anon 1852b). Similarly, the Staatliches Museum für Naturkunde Karlsruhe, which was begun as a royal collection for the Grand Duchy of Baden in the eighteenth century but which opened for public viewing by the early nineteenth century, was envisioned as a means of expressing commitment to a sense of homeland (see Anon 1845). Indeed, institutions such as the Paläontologisches Museum, Munich, and the Naturhistorische Museum, Dresden, were both opened during this era of expansion within the confederation of territories that would form a German state after 1871. The museum in Dresden was under the leadership of the distinguished naturalist Ludwig Reichenbach (1793–1879) from the 1830s as he led the development of the collection to reflect the order within the environment (Reichenbach 1836). The classification of German flora conducted by Reichenbach enabled a conception of the environment as belonging to a particular sense of locality and region (Reichenbach 1830). It is significant that the influential academic journal Natur und Museum was founded in 1871 and published in Frankfurt by the Senckenbergische Naturforschende Gesellschaft. This organisation was itself established in 1817 to pursue the study of flora and fauna but the dedication of the specialist publication in the era of German unification evidences the symbiotic relationship between the natural world and the socio-political world (Anon 1872). Natural history museums provided a means of assuaging anxiety or affirming identity within nineteenth century Germany as society was confronted with the challenges of the modern age.

A similar process can be observed with the displays of institutions within colonial possessions in North America, South America and Australasia (Sheets-Pyenson 1988a). Collections of specimens were used to display the connections of peripheral areas of European imperial control to the wider world (Sheets-Pyenson 1988b). This was the foundation of sites such as the Australian Museum in 1827 which was established by the Secretary of State for the Colonies, Henry Bathurst, 3rd Earl Bathurst (1762–1834) (Strahan 1979). This is also evident in the formation of Gesner's Museum of Natural History by the geologist Dr Abraham Gesner (1797–1864) in St. John, New Brunswick, Canada. Formed in 1842, this institution, which was exhibited to the public as a means of civic improvement, exhibited minerals, rocks, fossilised flora and fauna as well as contemporary specimens from the natural world (Anon 1842). Combining objects from New Brunswick and Nova Scotia with samples from Britain and the United States, this institution functioned to entwine the histories of the new world and the old as well as the imperial home and colonial possession. Elsewhere in Canada, museums of natural history were being founded during the same era. In Quebec City, Pierre Chasseur (1783–1842) established a public museum of natural history in the 1820s, a scheme which gave access to the study of the natural environment of the region which was also combined with a variety of other historical curios and paintings. Chasseur's work was regarded highly by the authorities of Lower Canada, who funded the museum's work with considerable investments at several points during the 1840s as profits from the initiative were limited (Buchanan and Wicksteed 1845: 554). This commitment by the state to ensure the continuation of natural history museums was also present in the support of the Montreal Natural History Society in 1841 as their premises were bought by the city council (Anon 1841: 506). This focus on natural history took place during the debates surrounding the 1840 Act of Union when Upper and Lower Canada were merged into a single dominion state. The use of natural history became a means of establishing a sense of place. In this manner, it is through the acquisition and display of natural history in ordered structures that political or cultural identity is forged or preserved.

Within the United States, public and private initiatives led to the creation of natural history museums in the nineteenth century which affirmed the character of the nation (after Conn 1998). Peale's Museum in Philadelphia, founded by Charles Willson Peale (1741–1827), emphasises this usage of environmental specimens of the past after its formation in 1801 (Figure 2.3). Within Peale's display, fossils and geological displays were included alongside paintings of the great heroes of the Republic. The elements of natural history were ordered along the categories defined by Linnaeus as one of the nation's first museums established the antiquity of the country as the inheritance of the people of the United States. Central to this design of the exhibitions was the reconstructed display of the fossilised bones of a mastodon (Peale 1803). These remains, discovered in upstate New York, were innovatively arranged to demonstrate the scale of the living creature, but they also provided a 'narrative of significance' for Philadelphia, Pennsylvania and the wider nation (Looby 1987). In this manner, rather than a 'young country', this was a land of renown and with a history equivalent to that of any European kingdom (see Semonin 2000). Such was the significance of the museum that the State of Pennsylvania sought to institutionalise the collection after the 1820s as declining revenues threatened the continuation of Peale's work and its moralising effect on civic life (Anon 1840). The display of natural history as a means of exerting American identity within the nation and across the world was also evident in the formation of the prominent Museum of Comparative Zoology at Harvard University in 1859 (Anon 1861). Supported by the Massachusetts State Legislature, the institution was founded by the scholar Louis Agassiz (1807–1873) as the centre of study for natural history (Anon 1864–1865). The museum's exhibitions were formed to highlight the variety of life within various epochs and it set itself apart from European collections through its distinctive arrangements of type (Anon 1871). The pioneering arrangement of display cabinets focused upon the distinctive areas of the world where species emerged thereby elevating the significance of 'American' natural history.

This exhibition type which focused on emphasising concurrent relationships was promoted elsewhere in the United States through the work of Henry Augustus Ward (1834–1906) who had studied with Louis Agassiz at Harvard University. As a Professor at the University of Rochester, Ward founded a company which provided models and specimens for colleges and institutions across the United States. This work was supported by extensive surveys of European collections which were found to be thorough but entirely absent of a means of understanding the context of natural history:

> Such a collection is of value if earnestly and thoroughly studied in connection with the strata that furnished it, but it is utterly insufficient to give a correct

idea of the broad features of ancient animal life at various times and over the entire globe (Ward 1870a: 2).

Ward also stated that the use of casts of specimens was vital in the role of education. The necessity of providing contextual data rendered 'authenticity' a secondary concern. Indeed, casts could be painted and fabricated to bear a close

Figure 2.3 Charles Willson Peale. The artist in his museum. Detroit Publishing Company Photograph Collection. Library of Congress, LC-D416-366.

resemblance to the original in texture and colour but what was important was the wider collection:

> None but those who have tried it, can fully understand the amazing difference between the fruitless effort at teaching Natural History without the *thing*, and in holding up before the pupil a visible, tangible illustration (Ward 1870b: vi).

This encouraged the formation of significant natural history museums across universities in the United States. For example, the Peabody Museum of Natural History at Yale University was founded in the 1860s from the donation of the financier George Peabody (1795–1869), which provided opportunities for furthering the study of this developing field (Anon 1866: 15). Similarly, the Hitchcock Ichnological Cabinet at Amherst College featuring fossilised tracks was displayed from the 1850s (Hitchcock 1865). The placing of specimens within this arrangement provided an understanding of the wider environment but it also developed as a means to mobilise natural history for the purposes of Agassiz's vision of humanity. As a promoter of 'simultaneous existence' over evolution, Agassiz argued for the importance of particular variations between different regions with regard to animal and plant life as well as humans (Agassiz 1866). A comparative vision provided a framework to grade human societies on higher and lower orders as the adaptation present within natural history was used for a racist classificatory system (Agassiz 1859). The orientation of exhibits towards a eugenicist agenda that affirmed separation and distinction between groups evidences the significance of museological display. Indeed, it was during the mid-nineteenth century that a form of public exhibition through the use of dioramas came to prominence as a means of placing visitors physically within the setting of the natural world of the past to affirm the significance of this history (Wonders 1993).

Natural history dioramas depicting the environment of flora and fauna were a marked shift from the arrangements of display cabinets which had dominated the understanding and perception of the environment (Conn 1998: 70). Visitors were now not obliged to draw upon the order and dynamism of the genealogy of the natural world but to regard the preserved stasis of reconstructions and fabrications. A synchronic vision of a past environment was presented rather than a narrative of development through chronology. The rapidity with which this mode of representation began to be used within natural history museums was impressive. Alongside dioramas, cosmoramas were also brought into usage, where the individual is placed into a panoramic vision of a scene as a means of providing immersive experiences for visitors. Peale's Museum in Philadelphia had used this means of entertainment and engagement in their exhibitions of portraiture in the 1830s. The branches of Peale's Museum in New York, Baltimore and Charleston also employed this device. In New York, Scudder's American Museum, which opened in 1810 in the city's former poor house in Downtown Manhattan, used large paintings of scenes to accompany the exhibits of taxidermy and objects of natural history (Scudder 1823). These examples were a means to elevate the visitor experiences, to ensure distinction for the private museums and to enable the institutions to remain profitable in a crowded marketplace for entertainments and distractions. In Sweden and Germany, these scenes began to be regarded as essential in the display of natural history to the public (Wonders 1993).

In Britain, the use of taxidermy in exhibitions was accompanied by the use of this new means of 'seeing' the display. This was pioneered with the creation of 'Mr. Bullock's London Museum and Pantherion' at the Egyptian Hall in Piccadilly, London, after 1812. William Bullock (1773–1849) had previously developed a collection of 'natural curiosities' for display in Liverpool (Bullock 1808). Using wax figures and preserved specimens, these displays provided dramatic reconstructions of the environment (Bullock 1817). However, in the new premises for the museum in the capital an engaging and immersive scene was developed:

> One department of the Museum (the Pantherion), completed with much labour and great expense, is entirely novel, and presents a scene altogether grand and interesting. Various animals . . . are exhibited as ranging in their native wilds and forests; whilst exact Models, both in figure and colour, give all the appearance of reality; the whole being assisted with a panoramic effect of distance and appropriate scenery . . . (Bullock 1812: iv).

Similarly, the Guiana Exhibition held during 1840 in the exhibition space known as the Cosmorama in 209 Regents Street, London, also provided an application of this technique of using preserved animals and panoramic vistas for natural history. This display was drawn from the collections of Robert Schomburgk (1804–1865), whose extensive expeditions to Guiana in the 1830s had brought back over 500 specimens of flora and fauna to exhibit (Schomburgk 1840). As a centrepiece, an extensive portrait of the environment of Guiana was presented with displayed animals, plants and recreations of the huts of the indigenous peoples of the region. As part of this exhibition, three individuals named as Corrienow, Saramang and Sororeng, from groups in Guiana, who had travelled with Schomburgk were present to display and explain items to visitors. This exercise in colonial power and perspective reinforced notions of race and time for early Victorian Britain (Coombes 1994). This powerful gaze which was formed through such arrangements served as a frame with which visitors could assess the wonders of the natural world as a static entity from a point removed; observers of natural history are not connected to the scenes they witness. This point of disconnection was significant – it rendered the forces of natural history into a governable and passive form. Such an effect was captured in the satirical work by Albert Richard Smith (1816–1860) in his treatise, The natural history of the idler upon town (Smith 1848). In this account, the visitor to the diorama or cosmorama in the city's museums was able to be in any corner of the world, past or present, but the individual was offered a means of avoiding any direct engagement with the reality of such places (Smith 1848: 103).

This sense of separation through looking upon the static elements of natural history was also fostered by the development of taxidermy within museums during the early nineteenth century (Lee 1820). The remains of animals which had been preserved and mounted for display had become a central feature of natural history museums for their ability to capture the immediacy of an environmental scene. The 'art' of taxidermy had been developed with the expeditions of the eighteenth century to Africa and Asia as a means of ensuring the survival of new finds (see Manesse 1787). The ability to preserve nature in this manner offered a commanding vision of the past and present where the dynamism of natural history was rendered a fait accompli. The spectacle that the arrangements of animals impressed upon the visitor was immediate. For example, Sir George William Lefevre (1798–1846), whilst travelling in Berlin, noted that the displays of the Museum für Naturkunde of woodpeckers arranged as if they were 'tapping the trees' had a 'very pleasing' effect (Lefevre 1843: 168). With significant collections of specimens in France and Germany, museums in Britain began developing taxidermy collections to provide visitors with an engaging, effecting response which also led to the further development of natural history within society (see Brown 1840). Indeed, through the increased use of taxidermy as a means of exhibiting, university, regional and private museums flourished during this era.

By the 1840s the Natural History Society of Manchester had a public museum, the Liverpool Royal Institute exhibited to citizens, and the Bristol Institution for the Advancement of Science and Art and the Newcastle-Upon-Tyne Natural History Society possessed display cases replete with geological specimens, fossils and taxidermy to inform, educate and entertain residents in the principal cities of industrial Britain (Swainson 1840: 74). Smaller institutions, such as the museum built for the Montrose Natural History and Antiquarian Society in Scotland, or the building constructed for the Penzance Natural History and Antiquarian Society in Cornwall, also utilised taxidermy to excite their patrons with the wonders of nature. Perhaps unsurprisingly, within this context, the Great Exhibition's display of natural history in 1851 was focused particularly on models and preserved specimens (Anon 1851b: 646). Such applications of preservation techniques were undertaken to ensure that the objects appeared life-like to the observer but were rendered immobile in reality. The British Museum increased its use of taxidermy within exhibitions during the 1840s and 1850s as it sought to develop displays which provided a 'natural' setting for visitors:

> The British Museum . . . in 1836 we find the natural history collections were as follows: Mammals, species 405; birds, species 2400; constituting altogether in specimens the sum total of 4659. Of reptiles we could boast — species 600, specimens 1300; fish 1000 specimens. . . . In 1848 an extraordinary increase (marking the great interest taken in taxidermical science) had taken place; we now had added to the British Museum since 1836, 29,595 specimens, comprising 5797 mammals, 13,414 birds, 4112 reptiles, 6272 fish (Browne 1878: 12).

Such an extensive collection was promoted as a point of national pride as the museum vied with rival collections in Paris, Berlin and Vienna for prominence in the field of natural history. As such, preserved animals were part of a wider dialogue regarding colonialism and nationalism. However, it is their arrangement and display within the museum that served to emphasise the perception of natural history as an object to conserve, to order and to divorce from social experience by rendering it the object of the visitor's gaze. This provided a means of capturing the environment to assert the same level of control within contemporary Victorian society (see Poliquin 2012). The Powell-Cotton Museum in Kent, which was founded by Percy Horace Gordon Powell-Cotton (1866–1940) in 1896, is an example of this combination of diorama and taxidermy. Large sections of the exhibition space were devoted to displays which set the visitor in an alternative environment. Similarly, the Booth Museum in Brighton, which opened in the 1870s as a repository for the collection of Edward Thomas Booth (1840–1890), also employed this technique in its exhibition as it sought to show how the collection of preserved British birds would have appeared in full flight (Booth 1901). In an era when urbanisation and industrialisation had altered established orders and when the environment was being reshaped through agriculture and commerce, the display of natural history as a static entity within the purview of the observer reinforced the social and political order (after Alberti 2011).

It was in this definition of refinement that the work of Charles Darwin (1809–1882) caused a furore. Whereas previous conceptions of natural history had assumed it to possess an immutable, fixed state, Darwin proposed a vision of this subject area that was in a constant state of development and adaptation. Indeed, what was examined in the geological or palaeontological evidence that was discovered and classified by scholars was not some stable picture of the past but a world in perpetual, sometimes gradual but at other times violent, alteration. Such perceptions of natural history had already been forwarded in the work of early nineteenth century geologists who had perceived within rock formations and stratigraphy an era of motion and change (Hutton 1788; Sedgewick 1821). The pioneer of this field was the British geologist Charles Lyell (1797–1875), who forwarded within his multivolume work 'Principles of Geology' (1832) an understanding of the world as a constantly changing system. Indeed, identifying and explaining 'change' was at the forefront of this work. Lyell's theory of 'uniformitarianism' argued that the forces of geological and climatological alteration that exist in the past also exist in the present. As such, the world is in a continuous process of modification; natural history is a narrative of movement and reformation. This contrasted with the view of the 'catastrophists' whose perception of geological formations as evidence of intense moments of upheaval in an otherwise stable world of tranquillity favoured stability over change (Whewell 1837). Lyell's (1832) assessment of constant adaptation demonstrated an environment in continual adaptation; stasis could not be perceived within the geological record and what appeared now to be inert could once have been dynamic and could return to being so in the future:

> Should we ever establish by unequivocal proofs, that certain agents have, at particular periods of past time, been more potent instruments of change over the entire surface of the earth than they now are, it will be more consistent with philosophical caution to presume, that after an interval of quiescence

they will recover their pristine vigour, than to regard them as worn out (Lyell 1832: 189–190).

It is on this definition of natural history in motion, that past and present activity in the environment could be defined as ongoing, that the theory of evolution was premised. Published in the two works, 'On the origin of species by means of natural selection' (Darwin 1859) and 'Descent of Man' (Darwin 1871), evolution is defined as a process of adaptation and change amongst flora and fauna (Figure 2.4). Through the examination of fossilised remains and studying present-day specimens, Darwin highlighted physical traits and behaviours which had formed through alterations

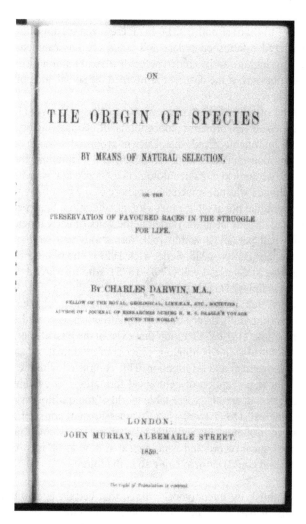

Figure 2.4 Title page of 'On the origin of species' (1859) by Charles Darwin. Library of Congress, Illus. in QH365.O2.

in the environment and by 'natural selection'. It is the latter which validates the significance of viewing natural history as a record of adaptation as it is within 'natural selection' that change is regarded as a necessity. It is through adaptation and development that survival is ensured:

> When under nature the conditions of life do change, variations and reversions of character probably do occur; but natural selection . . . will determine how far the new character thus arising shall be preserved (Darwin 1859: 15).

Evolutionary theory which is defined by the notion of change was regarded as a threat by scientific and religious figures as it positioned the wider environment in a permanently altered state. No longer was natural history reflective of eras of tranquillity which were only disrupted by cataclysmic episodes of geological, climatological or divine design, this was an account of a world in motion. This movement may well perhaps have been imperceptible at times but it was always occurring. Nature in a pristine state did not exist; indeed it could not exist as that primeval stage would have been required to adapt and develop. The revelation of change and development within humanity that Darwin provided challenged accepted norms by stating we are a physiological product of that alteration and a result of the contradiction and dynamism within natural history. The sense of stasis within the environment which had determined the relationship between human society and the world began to be replaced with a sense of movement.

The notion of natural history as a dynamic element in the past and the present was also fostered by the popularization of the field of study within the middle classes in Britain and the United States during the latter half of the nineteenth century. Geology, palaeontology and climatology, which had once been the preserve of the elite, became an interest of a wider section of the population as accounts of fossil discoveries, rock formations and expeditions to unexplored regions of the world were published. Indeed, from the 1850s to the 1900s a 'mania' for natural history could be observed within society. The coining of the term 'dinosauria' by the British biologist Sir Richard Owen (1804–1892) and the self-promotion of his findings catered for a growing public enthusiasm for knowledge and discovery (Owen 1841). This was reflected in the significant proportion of the Great Exhibition of 1851 in Hyde Park in London being devoted to the natural history of nations and peoples of the world (Official Catalogue of the Great Exhibition 1851). The display of flora and fauna was provided for entertainment but there was an educational and reforming element to these arrangements. Adults and children alike were encouraged to contemplate natural history and the adaptation of animals and plants to their environment as a means of rational improvement within themselves (see Anon 1851a). The use of natural history as an instructional device was also witnessed in the close association between the industrial and the natural within the displays. Rather than representing two distinct fields, the emphasis on originality and innovation ensured that the exhibitions of fossilised or preserved remains as a process of change were able to complement the exhibits of science and technology as a demonstration of development (Anon 1852a). Natural history was ascribed a status for the individual as a means of improvement but also for wider society to comprehend the vast changes that were occurring with the acceleration of the industrial revolution.

In this manner, the narrative of natural history was an account through which the development of nineteenth century society could be understood (see Anon 1829). Just as industry had developed to meet the demands of this new age, the record of the past was littered with the remains of fossilised flora and fauna that had failed to respond similarly. The pioneers of this field, such as Richard Owen, asserted this thesis of inadaptability in their assessment of the specimens they examined.

> No doubt the type-form of any species is that which is best adapted to the con-
> ditions under which such species at the time exists; and as long as those
> conditions remain unchanged, so long will the type remain; all varieties depart-
> ing therefrom being in the same ratio less adapted to the environing conditions
> of existence. But if those conditions change, then the variety of the species at
> an antecedent date and state of things may become the type-form of the species
> at a later date, and in an altered state of things (Owen 1866: xxxv).

Drawing upon Darwin's published works, natural history could now be regarded as a testimony of failed endeavour and successful innovation. This was a mirror of late nineteenth century society which would appear to affirm the values of commerce as a uniform law. The dynamism within this past was used to comprehend the present. This was also seen in the model dinosaurs sculptured by Benjamin Waterhouse Hawkins (1807–1894) for the new site of the Crystal Palace Exhibition Hall in South London. Operating under instruction by Richard Owen, these replicas were built according to the latest research but their effect was to excite visitors with their scale and ferocity whilst indicating the moral, political and technological lessons of this encounter. The physical presence of natural history in this dioramic scene evidenced the necessity of development, the constancy of progress. Without this advancement, society would be seemingly doomed to the same fate as their giant prehistoric predecessors whose fossilised bones, displayed for public amusement, were all that remained for all their colossal scale.

The popularity of these displays and the need to preserve and display the collections housed at the British Museum led Richard Owen, as Superintendent of the natural history departments of the British Museum, to lead the calls for a separate museum space (Owen 1862). By the 1870s, work had begun on the site in West London which, alongside the South Kensington Museum (originally the Museum of Manufacture when constructed in 1852 and later renamed the Victoria and Albert Museum), made up the museum quarter which developed after the Great Exhibition. The museum, designed by Alfred Waterhouse (1830–1905) in a Romanesque style, was opened in 1881 as the British Museum (Natural History) (Figure 2.5). Within the new galleries, visitors were informed through the arrangement of specimens of the way in which nature had depended on alteration and improvement to adapt to altered environmental conditions (Günther 1885). In the geological and palaeontological galleries, guidebooks reminded those who toured

Figure 2.5 Natural History Museum, London.

the exhibits and displays that they were observing the ever-changing aspects of nature as they examined the objects:

> If in a similar manner we investigate those larger layers of Chalk and Limestone, Sandstone, Clay, or Slate, composing the Earth's crust, we not

only find that they rest upon one another, so that we can judge of their relative age by the order of their superposition, but that, like the layers of soil below London they are often full of relics which tell of the former inhabitants that lived, flourished, and died out, to be succeeded by another race which have in their turn shared the same fate (British Museum (Natural History) 1886: 2).

The museological display of natural history at the museum affirmed the status of the study as a pursuit of distinction as it catered to the educated social classes but it also ensured the perception of natural history as characterised by development and change. It is this aspect of the study which contributed to the public fascination for geology and palaeontology in the late nineteenth century. Far from collecting materials that represented stability, rocks and fossils appeared to show the ever-altering nature of the world and humanity's place within this system (Taylor 1885).

The American Museum of Natural History in New York was founded during the same era in 1869 by popular appeal amongst the city's political elites who were eager to have a museum of natural history to reflect the 'metropolitan' status (Anon 1870) (Figure 2.6). Originally housed in the Arsenal in Central Park, the grand, classical, purpose-built premises on Central Park West were completed in the late 1870s. At the dedication ceremony for this new site, the President of Harvard

Figure 2.6 American Museum of Natural History, New York. Detroit Publishing Co. Library of Congress, LC-D4-14285.

University, C.W Eliot (1834–1926), opened his address with an assertion of the value of the institution for the city and the nation:

> In whose honour are the chief personages of the nation, State, and city, here assembled? Whose palace is this? What divinity is worshiped in this place? We are assembled here to own with gratitude the beneficent power of natural science; to praise and thank its votaries, and to dedicate this splendid structure to its service. The power to which we here do homage is the accumulated intelligence of our race applied generation after generation to the study of Nature; and this palace is the storehouse of the elaborated materials which that intelligence has garnered, ordered, and illuminated (Eliot 1870: 49).

However, in this same speech Eliot further defined the purpose of natural history being threefold: the development of an inquiring mind amongst humanity, the assertion of the inevitability of progress and the necessity for implementing the 'doctrine of hereditary transmission' (Eliot 1870). In this manner, natural history was used to assert the need for a programme of eugenics. The utilisation of the geological and palaeontological evidence to elucidate a notion of 'social Darwinism' had been pioneered by Herbert Spencer (1820–1903), whose work had moved evolutionary theory beyond biology into politics and society (Spencer 1864). Similarly, Thomas Henry Huxley (1825–1895) applied Darwinian theory to consider current and historical specimens of humanity for 'advanced' and 'regressive' traits (Huxley 1863). With the work of Sir Francis Galton (1822–1911), who developed the 'science' of eugenics during the nineteenth century, fossils and specimens were used to define racial profiles and assert the superiority of groups of people over others (Galton 1869). Museums of natural history could become knowing or unwitting communicators of these principles as their emphasis on development and adaptation enabled the projection of this vision of humanity. Indeed, the American Museum of Natural History became a centre for eugenics study during the early twentieth century as Henry Fairfield Osborn (1857–1935), initially as curator and then as president, utilised the collections to assert his belief in a hierarchy of races (see Osborn 1916). In 1921, the museum hosted the Second International Eugenics Congress which also included a specially convened 'Eugenics Exhibition'. This extensive display was divided into two sections, an exhibition on the palaeontology of man prepared by the museum in the Hall of the Age of Man on the fourth floor and a display on eugenics in eighteen alcoves in the Forestry and Darwin Halls on the first floor. Natural history as an account of development and progress had been mobilised to assert the primacy of 'European' races and the necessity of preserving this status (Laughlin 1923).

Dynamism and development within natural history

Within the United States, the use of dioramas, featuring casts, models or taxidermy, became prominent after the 1880s as new museums were founded through the availability of objects and specimens. This utilisation of new museological practice could be regarded as a response to the showmanship and spectacle of P.T. Barnum's (1810–1891) exhibitions. After purchasing Scudder's American Museum, the natural history displays were developed to include specimens of natural 'curiosities' as well as fabrications designed to provoke visitors' sense of order and rationality (Durst and Barnum 1849). From 1841 to 1865, Barnum's American Museum displayed fraudulent exhibits which appeared to subvert the established knowledge of classification. Despite the obvious appeal to sensationalism and scandal, Barnum's American Museum mirrored the work of its contemporaries within the museum world. Barnum's arranged natural history as an object of perusal through the use of dioramas. Established, professional institutions may have avoided the heavily-commercialised and popularist displays favoured by Barnum, but nevertheless they shared the same effect of depicting natural history as an object of the visitors' gaze. Perhaps Barnum's Museum was at a very different end of the spectrum, but it still reinforced the use and function of the visual display in institutions. This can be observed in the camel caravan diorama displayed as the centrepiece of the American Museum of Natural History after its opening. Entitled 'Arab Courier Attacked by Lions', it was modelled in 1867 by the French botanist Jules Verreaux (1807–1873) and brought to the United States for exhibiting in 1869. For Barnum's American Museum as well as the American Museum of Natural History, what was significant was the concentration on preservation for observation.

In the larger, venerable institutions of natural history the vision of a fixed and stable notion of natural history also affirmed the wider changes within society, politics and culture. One of the first museums to pioneer the use of alternative modes of display was the Milwaukee Public Museum which, under the guidance of the taxidermist and naturalist Carl Akeley (1864–1926), developed an extensive collection of preserved flora and fauna. The Milwaukee Public Museum responded to public and scholarly interest by presenting nearly 2000 fossils and nearly 1000 mounted animals. However, it was Akeley's dedication to scientific rigour that drove his work in the field. This was a movement away from the anthropomorphic scenes in which the specimens were placed in varying levels of unnatural or absurd poses to one which emphasises context and experience (Anon 1883). This use of a diorama and taxidermy to inform individuals but also to reform society and reiterate social norms was evident in these displays. Whilst such engaging exhibitions provided a steady stream of visitors, the reconstruction of environmental scenes replete with animals and plant life placed nature as a subject which was certainly not ongoing and affected by human interaction. Akeley also worked with other museums across the United States to form engaging exhibitions and displays. Indeed, his work in Chicago laid the foundations for the development of the renowned collections within The Field Museum.

This institution was born out of the specimens which were drawn together for the Chicago World's Fair in 1893. Examples of taxidermy and fossils were displayed to visitors at the site and the popularity for the dynamic display led to the formation of a museum for these pieces for the purposes of civic improvement (Anon 1894). Originally named the Columbian Museum of Chicago in 1893, it was changed to the Field Columbian Museum in 1894 before becoming The Field

Museum in 1905. The institution took its name from the wealthy businessman Marshall Field (1834–1906) who had been persuaded to fund the institution which was located in one of the few buildings of the World's Fair that was not demolished. After 1894, the museum became increasingly reliant on the use of dioramas as a means of engaging visitors as elaborate scenes were painted and specimens arranged to provide colourful yet static images of the natural world (Metzler 2008). These displays were valued for their presumed sense of permanency as if the technique of taxidermy had captured the perfect form of nature which was now to be exhibited (Dickerson 1914). The success of the displays in The Field Museum provided Akeley with another position at the outset of the twentieth century at the American Museum of Natural History. In this role, further tableaus were arranged of specimens which provided a 'natural' setting for visitors to examine fauna and flora. Such depictions arrested the environment for the purposes of removing contemporary society from such relationships. In this regard, natural history was an object of vision not a scene in which visitors participated.

This removal of society from the environment through the use of dioramas provided an understanding of the natural world which was focused on use and exploitation. Nature as a pristine setting could be exhibited in a museum whilst the power conveyed in the presentation of discrete vistas of animal and plant life communicated the sense of ownership (Landes, Young and Youngquist 2012). As such, the sponsorship of these natural history institutions in the United States and Britain by wealthy industrialists was entirely fitting. This was a means of establishing possession. However, as the use of dioramas and the display of taxidermy secured an image of nature for viewers it became important in the growing concern for conservation. With the threats posed to the natural environment due to the exploitation of resources, the natural history museum became a site of sanctuary for vulnerable habitats. Therefore, to represent the flora and fauna of a region or nation was an act of guardianship rather than possession. Such an alteration in the relationship towards natural history is reflected in the museums of the early twentieth century. The National Museum of Natural History in Washington, D.C. was constructed in 1910 after the initial display of the collections in the mid-nineteenth century in the Smithsonian Institution Building (see Rathburn 1913). The focus within this institution was to provide visitors with a focus on the contemporary environment in order that the effects of development elsewhere across the nation were ameliorated (United States National Museum 1919). Similarly, at the outset of the twentieth century, the American Museum of Natural History in New York was also considering orientating itself towards a concern for conservation in its exhibition (Osborn 1910). In an era of change and development, when the lives of individuals were being radically transformed, the museum of natural history was a place of stasis (Hutchinson 1897; Lucas 1901).

This is also seen in the exhibitions and displays of the Horniman Museum, London, which opened in 1901 for the 'recreation, instruction and enjoyment' of citizens. Built with the collections of Frederick John Horniman (1835–1906), the central feature of the new museum was the taxidermy that provided a vivid image of the natural world (Anon 1901). Extensive arrangements of preserved animals were arranged in the North Hall of the institution on the wall-cases and centre-cases of the exhibition room. To accompany this arrangement, a large display of a 'Survey of the Animal Kingdom' and 'The History of Animals' was provided to visitors as a means of placing contemporary human society at the apex of this system (Anon 1904). Under the supervision of the anthropologist A.C. Haddon (1855–1940), the museum was organised as a demonstration of the evolutionary processes which ultimately formed the stable present. This arrangement was maintained in place during the early twentieth century with only minor adaptations as the value of natural history for education and improvement was stressed by the museum. Where the displays were altered after a decade of the institution's opening, it was in the use of models and fossils to highlight convergent and divergent evolutionary models. The abilities of species to swim, creep, burrow, run, parachute, fly and climb were used as points of classification as the forms of 'Animal Locomotion' were demonstrated (Anon 1912). The alteration of the exhibition content provided a point of comparison with the modern world as mass transportation systems and the widespread proliferation of the motor car enabled the defining character of the modern age: speed. Therefore, within this display of the natural world, the acceleration of human society was given a precedent.

This process of appropriating the form and function of the natural world to explain contemporary phenomena can be observed in the response to the development of military technology during the First World War (1914–1918). On the battlefields of the Western Front in France and Flanders, soldiers and commentators from all combatant nations regarded artillery, machine guns, Flammenwerfers, mines and tanks through metaphors and allusions to aspects of natural history. It was the tanks, introduced by the British Army in 1916, that appeared particularly redolent of objects witnessed within the display cases of natural history museums:

> A tank. . . . What a name for this marvel, this weird, weary dinosaur whose bowels are racketing engines, whose teeth are machine guns, and whose hide is two inches of proof steel, which can climb over cliffs and into valleys, and is impervious to all the weaknesses of wheeled things (Holland 1917: 691).

> Man alive, it's a Tank! Wobbling and swaying, dipping and rising, like a dazed Dinosaur, from the shell wallows, and spitting dragon flames, 'Old Faithful' is carrying out orders to the letter attacking a whole German Army with superb but ridiculous assurance. . . . Big crumps are falling perilously near our friendly mammoth of the stone age, materialized out of the dreams of romances to help us purge the earth of the unclean Hun . . . (Anon 1916a: 464).

In the confrontation with the industrialised warfare on the battlefields and the use of technology to inflict dismemberment and death upon soldiers, the figures of natural history served as a means to engage with the forces of the modern world (Fussell 1975; Winter 1992). Indeed, within wider popular culture during the conflict, allusions to natural history served as a means to mobilise opinion either to support the war or to oppose its destruction. In effect, natural history offered a means of legitimising the militarisation of society through a comparison with the adaptations in fauna and flora within the environment. Indeed, United States Solicitor General James Beck (1861–1936) spoke with reference to the necessity of preparation for the war by reference to the nation's role in producing this new technology of war:

> The steamship, the telegraph, the cable, the submarine, the telephone, the machine gun, the aeroplane, and, it is claimed, even the steel-sheathed motor trucks, which, like gigantic Dinosaurs, have only recently moved across the battlefield, are all the product of American inventors (Beck 1917: 87).

Conversely, references to natural history were also used to deter the entry of the United States into the war. The Anti-Preparedness Committee, comprised of pacifists, socialists, religious figures and artists and which grew out of anti-war sentiment within New York, constructed a fifteen-foot model of a *Stegosaurus* (named 'Jingo'), mounted and displayed as if it was a museum exhibit, with the explanatory label 'All Armor Plate – No Brains' (Anon 1916b: 2). This display was toured across major cities in the United States as an object through which the failed endeavours of militarism could be shown. In this exhibition, just as the dinosaurs had appeared to increase in size and defensive capabilities before succumbing to extinction, so too would the war-mongers of the world (Anon 1916c). The malleability of the past environment could be observed in this instance. Indeed, whilst questioning the accuracy of the models, Dr William D. Matthew (1871–1930), curator at the American Museum of Natural History, remarked that the use of this model for processional and moral purposes demonstrated the success of his institution in informing and educating the public (Matthew 1916). With the advent of the First World War, natural history had been mobilised.

The museums of natural history in combatant countries played their part in the war effort (see Kavanagh 1994). In Britain, the call to use the collections at the disposal of museums across the country was made by a curator at the Natural History Museum, Francis Arthur Bather (1863–1934). He declared it a patriotic duty to use the exhibitions and displays to demonstrate the virtues of the nation, build international alliances and further the pursuit of victory (Bather 1915). Therefore, the natural history museums played a significant part in the war effort as the repositories of knowledge and as the records of environmental adaptations which could provide important insights into developing new technologies or innovative processes:

> As conservators and curators by profession it is our task always to preserve the past, not merely because it is past, but because what in its day was good may in the Whirligig of time again come to its own (Bather 1914: 250).

After 1914, a number of museums began providing galleries, tours and special exhibits for soldiers and the wider public as a means of demonstrating the utility of natural history for winning the war. With the threat of museum closures due to wartime austerity in 1916, it was the Natural History Museum and its display of fossils, fauna and flora which was maintained because of its role in entertaining 'colonial' troops, maintaining public morale and providing vital information for pursuing victory:

> It is the knowledge of the life-history of the micro-lepidoptera which has proved of such incalculable service to this country during the recent wars in preventing their ravages among our food supplies, ravages which ... would have proved as disastrous as the worst engagement yet fought (Anon 1916d: 83).

Whilst natural history museums became another arm of service, the collections in museums across Britain could be framed in relation to the conflict. Specialist talks and exhibitions such as 'Evolution and the War', 'Geology and the War' or 'Biology and the War' became regular features of the Natural History Museum in London and in smaller regional museums. Similarly, the American Museum of Natural History provided a special discussion on minerology and geology during the war to increase public support and foster scientific development for the war.

> A small exhibit illustrating the relation of minerals to the production of munitions of war has been installed in two cases ... in the Hall of Minerals. This series, the assembling of which was begun in June, aims to visualize the steps in the development of war munitions from the ore to the finished product and to emphasize the need of establishing an adequate domestic source of supply of the ores of the rarer metals (Whitlock 1918: 61).

Similar processes were recorded in Germany within institutions such as the Naturmuseum Senckenberg in Frankfurt or the Museum für Naturkunde in Berlin (Anon 1915). Whilst the war disrupted the usual operation of the institutions, exhibitions were provided for soldiers and the wider public to engage Wilhelmine German society with the role of natural history and the war effort (Anon 1918). The significance of this use of the environment is the malleability which fossils, flora and fauna seemingly possess that enables a shift in their use within institutions. Natural history was redefined during the conflict to support national identity and to further the war effort. In effect, this both reflected the previous usage of this heritage and set the agenda for how natural history would be regarded throughout the twentieth century. When faced with the hazards and threats posed by the modern world, natural history has served as a frame through which reassurance or reorientation has been sought.

The proliferation of museum guides and touring books during the interwar years in both Britain and the United States would appear to affirm the value of natural history in reforming and consoling society as the popularity of this subject grew in a politically and economically turbulent era. For the natural history museums of the 1920s and 1930s, the focus on order and stability within representations was present but so too was the sense of entertainment and excitement (Conn 1998).

Museums arranged group displays of fossils and taxidermy as a means of exciting the interests of the visitor (Lucas 1921). These displays were also accompanied by a greater sense of social responsibility expressed by natural history museums which stemmed from their wartime roles (see Evermann 1918). For example, the Field Museum of Natural History in Chicago began a programme of engagement with local schools and youth groups within the wider metropolitan area before the war but significantly expanded their operations in the 1920s. This work involved developing museum cases, filled with objects and specimens, which would be loaned out to groups to enable teachers and youth group leaders to build sessions around these pieces of evidence (Field Museum of Natural History 1928). Similar work was being undertaken by other natural history museums across the United States as outreach and access initiatives formed a means of providing social welfare. The Bell Museum of Natural History at the University of Minnesota (Roberts 1922) and the Milwaukee Public Museum (Thal 1921) both examined the potential of using their collections for educational purposes. The American Museum of Natural History began a schools programme in the 1920s as a means of improving the lives and characters of citizens (Sherwood 1920). The Director of the Museum, Henry Fairfield Osborn, heralded this work as demonstrating the capacity of public good that such institutions could have in society:

> The new spirit within the natural history museum is the educational spirit, and this is animated by what may be called its ethical sense, its sense of public duty, its realization that the general intelligence and welfare of the people are the prime reasons for its existence, that exploration, research, exhibition, and publication should all contribute to these ends, that to serve a community the Museum must reach out to all parts of nature and must master what nature has to show and to teach (Osborn 1920: 5).

Before and after the onset of the Great Depression, this work provided a means of both cultural inclusion and social engineering as natural history museums were regarded as a means by which society could be transformed. This can be regarded as an international movement, as institutions in North America and Europe all altered their exhibitions and displays to enable this new era of engagement and education. In Brussels, Professor Victor Van Straelen (1889–1964) had taken over as Director of the Musée royal d'histoire naturelle de Belgique after 1925 and had begun to make education a key feature of the institute's work. In Berlin's Museum für Naturkunde, the exhibitions were reordered under the direction of Professor Carl Zimmer (1873–1950) to provide a more engaging experience for visitors (Zimmer 1931). Indeed, this can be acutely observed in the Museum für Naturkunde und Vorgeschichte in Dessau, which was opened to the public in 1927. This institution was remodelled by members of the Bauhaus architectural school which had removed to Dessau after their initial location in Weimar (Rössler 2014: 62). Under their influence, the museum was designed for the efficient delivery of information to facilitate the use of the collection by visitors (Anon 1927). In his assessment of the state of natural history museums across the world for the thriving, colonial era Bombay Natural History Society (founded in 1883), the Anglo-Indian naturalist Stanley Henry Prater (1890–1960) regarded such developments as vital for the modern world:

> The object of the museums is therefore to lead the people to better ideals and to offer them new view-points in a way which they can understand and appreciate. The appeal is made more to the senses than to the intellect but it is made in such a manner that in addition to mere attractiveness it offers food for thought which is the first step to a higher and more intellectual ideal (Prater 1928: 533).

With museums now focused on access and engagement, the function of institutions had been altered by these early twentieth century developments but the emphasis on shaping public attitudes through natural history remained a vital aspect of exhibitions and displays. This status within society was called upon during the Second World War (1939–1945), when all aspects and institutions of the nation state were mobilised once again to instil commitment and maintain morale within the public. Natural history museums in Europe and North America used their collections to legitimise the conflict and to promote its continuance amongst the populace. From the national to the regional, exhibitions and displays were used for the war effort. In Britain, the Natural History Museum in London served as an important location for demonstrating the national 'spirit' throughout the war and as a centre of knowledge and guidance on the natural world for military purposes (see Haldane 1941). After the entry of the United States into the war in 1941, institutions began ensuring that their work concurred with wartime agendas (see American Museum of Natural History 1941). For example, the Illinois State Museum, which was founded in Springfield in 1887, provided visitors with a 'War Room' in their 'Youth Section' to inform individuals and communities with information on the pursuit of the war and its relation to natural history. The Director of the Museum, John C. McGregor (1881–1975), saw a direct correlation between the work of the institution and supporting the nation's drive for victory:

> In time of war the greatest single service which a museum may render the public is to show the natural history and art, the inspiration as well as the economic way of life which we are attempting to assure (McGregor 1943: 23).

This was further reflected in exhibitions of natural history which affirmed the righteousness of the cause. As Editor of the in-house museum publication, the author Virginia Eifert (1911–1966) oversaw the publication of a variety of pieces on exhibitions which emphasised the necessity of conflict as well as the restorative powers of the natural world:

> Today the Museum's exhibit of South Pacific birds is peculiarly interesting for its significance in gaining an understanding of today's story of man and his ways. They represent the birds which were blasted to bits in sky, sea, and

land-battle, and stand as symbols of a peace which one day will come again. They represent heroic Wake, Midway, and Laysan, and the other embattled islands of the Pacific, whose sacrifice in blood and birds is only a part of the great sacrifice taking place throughout the world (Anon 1942: 39).

The national parks of the United States were also mobilised for war as the natural history of these sites served as an emblem of American identity whilst the resources available for use were presented as under threat from the effects of wartime needs. In this manner, the displays of preserved animals, natural scenes and reconstructed habitats in museums provided evidence of a museum's ability to conserve the environment and inform wider society. The uses of natural history museums for the purposes of war was also demonstrated within the institutions which operated under the control of the Third Reich. Indeed, even before the advent of the war, there is a shift in the values and uses of natural history (see Epstein 2010: 248; Hutterer 2014: 28). It is during this era that the periodical *Natur und Museum* altered its title to *Natur und Volk*. After 1939, museums in Germany as well as annexed Austria and occupied France and Poland became tools through which Nazi propaganda regarding place, lineage and culture could be communicated (see Mackensen 1943). The Naturhistorisches Museum in Vienna contributed to this programme through the work of the anthropologist Josef Wastl (1892–1968). Under his supervision, the museum conducted racial studies of peoples and exhibited the results as part of an assessment of 'nature' within central Europe (see Wastl and Sittenberger 1941). The function of natural history was to provide the basis for a racially pure society and to assert the vision of the future Germanic state forwarded by the Nazi Party. Natural history served to emphasise the earliest foundations of the state but also the direction of society. In this manner, natural history reflects the basis of modernity; an imperious desire to impose values and ideals upon an imminent era beyond the confines of the present (after Giddens 1990).

Redefining society through natural history

The use of natural history museums and natural heritage sites during the course of the early twentieth century has cemented their position within the post-1945 world. Industry, conflict, economic instability, social change and cultural developments have been the forces of change which have shaped the understanding, use and representation of environmental history (see Bates 1950). It is within the museums, parks and gardens that a concept of natural history is developed and disseminated. These are places of education, mobilisation, action, assertion and reiteration as 'natural history', what appears to be a fundamental, universal aspect of our environment, has been moulded to suit social desires, anxieties, fears and aspirations. This is especially the case with representations of natural history in the latter half of the twentieth century, where heritage sites have become the locales for altering humanity's relationship with the environment by changing the past to inform the present and shape the future (see Saunders 1952). As western society was exposed to the dangers of a new age, natural history has been used to ameliorate, manage and dispel the risks faced by the individuals, communities and nation states (after Beck 1992). The representation of natural history is a means of consolation and isolation as the precarious nature of humanity's existence is justified and explained through a process of foreshadowing (after Bernstein 1994). Since 1945, western society has been encouraged to think through natural history to consider the present and the future. This stands in direct contrast to how natural history had been used throughout the course of the eighteenth and nineteenth centuries to assess the relationship with the past. The modern world is formed in relation to natural heritage.

This particular connection was initially relayed in the response by museums, parks and gardens to the threat posed by nuclear war after the end of the Second World War. In the shadow of the atomic arms race, natural history formed a means by which the wider populace could be informed and assured of the purpose and use of this technology. In 1948, the American Museum of Natural History was used as the site for educating society on the value of nuclear energy:

> The Museum acted as host to the Atomic Energy Commission in presenting, for the first time, a public exhibition prepared and operated by that organization, for the purpose of demonstrating the basic principles of nuclear fission. It emphasized some of the possibilities for peacetime uses (American Museum of Natural History 1948: 5).

In effect, this provided a means of 'naturalisation' for nuclear technology. To further emphasise this process, the United States Atomic Energy Commission donated thirty-five uranium ores from locations across the country to the 'nation's collections' in the 1950s (United States National Museum 1959: 173). Furthermore, the Museum Division of the Oak Ridge Institute of Nuclear Studies prepared a travelling exhibition during the 1950s and 1960s entitled 'You and the Atom' which toured expositions, fairs, galleries and other institutions. Such displays served to reassure and to calm wider society. Indeed, the American Museum of Natural History forwarded itself as a centre of calm in the maelstrom of this new world:

> The part that natural history museums can play – in showing how life exists in all parts of the world – can be a most useful one. It is a current fear that, with the possible use of hydrogen and atomic weapons, large cities may some day be removed from the face of the earth, but we could never go on with our work if we believed that this was inevitable (White 1953: 4–5).

However, it was in the context of the natural history museum that this display of cutting edge atomic science was rendered 'normal'. Surrounded by the displays of fossils and minerals, this innovation in technology acquired a 'natural' status. This can be seen in the display in the Denver Museum of Natural History, Colorado, during 1960. The exhibition was arranged in the Dinosaur Hall where the fossils and displays reinforced the notion that this technology was a part of the natural world. The presentation consisted of panels depicting the significance of nuclear power for industry, medicine, agriculture and the generation of power for home and factories. The utilitarian aspects of nuclear power were also displayed as this 'great new force' was shown to be part of the natural environment. This was further reinforced with a display examining the production of nuclear fuel, from its formation in the earth's crust, its extraction in mining operations and the eventual processing for use within power stations (Murphy 1960: 17).

As the anti-nuclear movement gathered support, sites of natural history were also used to demonstrate the effects of atomic weapons and nuclear fallout on the wider population. Indeed, the protest against this technology was fostered within museums of natural history. The influential Greater St. Louis Citizens' Committee for Nuclear Information, which was initiated in the 1950s to raise awareness of the effects of radiation, initially operated out of a room within the Museum of Science and Natural History in St. Louis, Missouri (Anon 1960). This organisation began lobbying for greater protection for civilians and conducted an extensive study of children's milk teeth for the presence of the radioactive isotope, Strontium-90, in the 1960s. This product of nuclear weapons testing was found to be increasingly present in the teeth of age groups born after 1945. Such results encouraged the St. Louis Committee to alter its journal name from *Nuclear Information* to *Science and Man* and eventually *Environment*. The situation of the 'Milk Tooth Survey' within the natural history museum provided a setting for the nuclear age to be depicted as an aberration within humanity's history. As the effects of a nuclear war became apparent, museums began using exhibitions and displays to emphasise the destruction which could be experienced through this man-made catastrophe. During the 1950s and 1960s, natural history was increasingly placed as the counterpoint to humanity's failings. In the Cleveland Museum of Natural History's exhibition during this era, after the display of animal evolution and adaptation the visitor was presented with a mirror into which their reflection was presented back with an information panel which stated that they were staring at the most violent and dangerous animal on the planet (Wells 1960: 364). Whereas once natural history glorified the achievements of society, it now served to indict its seemingly self-destructive tendencies. As threats and fears were confronted, natural history provided a point of comparison and potentially a means of survival into the future.

The assertion of natural history in the face of change has been a defining feature of the latter half of the twentieth century. However, this has largely been conducted within specific reference to conservation, ecological preservation and environmental issues. As the threat posed to the environment by industry, technology, agriculture and over-population was realised from the 1960s onwards, the sites representing natural history have taken prominent positions to alter perceptions of the past, present and the future. This was partially led by the publication of Rachel Carson's (1907–1964) highly influential work on environmental degradation brought about by intensive agriculture, 'Silent Spring' (1962). In this analysis, Carson collaborated with and referenced several leading figures within American natural history museums to demonstrate the impact of chemical pesticides on ecological systems and the health of human populations. This work affirmed a wider movement within the United States that saw institutions seeking to represent an 'ecological model' through their exhibitions and displays. This sought to place 'man' not as an owner or cultivator of his environment but as part of that wider ecosphere; if humanity was to face the challenges of a changing planet then it had to be undertaken with an understanding that humans were components of that natural world (see Parr 1959). This was first explored with the opening in 1951 of the Warburg Hall of Ecology in the American Museum of Natural History. This particular exhibition space focused solely on the environment of Dutchess County, New York. This specific place was not presented in a particular era but rather as a scene of interaction between humans, animals, plant life, water, earth, the seasons and the sun. Through a combination of scenery, objects and dioramas, a sense of unity in nature was provided to museum visitors. The exhibition hall was regarded by the museum as a new means to engage public consciousness for humanity's relationship to the environment. The 'complete' narrative which was presented was classed as a 'radical departure from the hitherto accepted method of museum exhibition' (American Museum of Natural History 1953: 28).

The provision of an 'ecological perspective' within natural history museums became a standard feature of many major American and European exhibitions over the course of the 1950s and 1960s. The Naturhistorisk Museum in Aarhus, Denmark, made this ecological perspective a defining feature of both exhibitions and research during this era (Thamdrup 1947). This work was supported by the studies undertaken in the Mols Laboratory, a field station where the analysis of natural systems provided insight into the function of environments (Thamdrup 1978). The alteration in this process is distinctive; visitors were now asked not to regard nature but to regard themselves in nature. The rise in environmentalism during this period provided a further definition to this new approach as the drive to encourage responsibility in conservation and preservation became essential components of sites of natural heritage. In Britain, this was fostered by the develop-ment of the National Parks which were formed by the Labour Government with the National Parks and Access to the Countryside Act 1949. At the centre of this initiative was a desire to conserve and protect:

> An Act to make provision for National Parks and the establishment of a National Parks Commission; to confer on the Nature Conservancy and local authorities powers for the establishment and maintenance of nature reserves; to make further provision for the recording, creation, maintenance and improvement of public paths and for securing access to open country, and to amend the law relating to rights of way; to confer further powers for preserving and enhancing natural beauty . . . (National Parks and Access to the Countryside Act 1949).

The areas which were selected for national park status during the 1950s under this legislation stood in contrast to the sweeping wildernesses preserved in the United States under the National Park Service Act of 1916. Within the 1949 Act, national parks in Britain encompassed the natural landscape as well as the people and places within them. Preserved for the nation, these were sites where natural history and human society were regarded alongside one another. However, the effect of this designation served the same purpose as the natural heritage was used to transform the lives of contemporary society and preserve a standard of living for new generations. The restriction of human settlement from areas with 'pristine' environments during the 1960s in the United States was undertaken for the same effect. The development of the National Wilderness Preservation System (NWPS) in 1964 after years of wrangling provided sites not of exclusion but of moral and economic improvement. Indeed, Congress defined the NWPS as serving the 'permanent good of the whole people' (Anon 1964). Natural history, whether in museums or national parks, was preserved and presented as a means of limiting and uplifting humanity's effect on the world.

This particular representation of the natural world was also present in exhibitions and displays of the histories of indigenous peoples during this period (Levine 1975). Within the Royal Ontario Museum, Canada, the 1969 exhibition entitled 'The Arts of Forgotten Peoples' stressed the successful engagement with the natural world within the First Nations. This was made in stark contrast to contemporary western society's increasingly troubled relationship with the environment (Rogers 1969). This trend towards ecological approaches is also reflected within the changing name of the publication of the Naturhistorischen Gesellschaft Nürnberg, which had given its title *Mitteilungen und Jahresbericht* the prefix of *Natur und Mensch* in 1970. It was during this era that 'ecomuseums' came to prominence within Europe as communities sought to preserve and sustain particular modes of living (see Davis 1999). Under the pioneering work of the French museologist Georges-Henri Riviere (1897–1985), experimental sites were developed within national parks where participants carried out traditional farming, crafts and land management (Riviere 1989). At these sites, the relationship between communities and the environment was displayed as key to understanding the past, addressing concerns in the modern world and ensuring a future for humanity. The rise in ecomuseums, across France initially but then within other European countries, reflected the challenges faced by cultural heritage sites in addressing environmental challenges of the era. Natural heritage was presented not as an object to observe but an event in which we participate.

This moral, political and social role within representations of natural history was further defined in the 1970s, when exhibitions and displays within museums focused upon the dangers of man-made environmental change. This was explicitly addressed by Chicago's Field Museum of Natural History with their exhibition held in 1974 entitled 'Man in his Environment'. This permanent display comprised 8,000 square feet of presentation area, two movie screens for public information shows and four three-dimensional scenes, as a major effort was made by the institution to engage visitors with the effects of humanity's expansion across the world and the exploitation of resources. What marked the exhibition was its insistence upon reflection:

> The exhibit deals first with natural systems and second with man's impact on them – leading to the inescapable conclusion that man is not the independent

master he so easily assumes himself to be, but one of earth's creatures, as dependent upon the environment as any other creature. Man in His Environment takes a global view of the most serious problems now confronting all man-kind and asks visitors to involve themselves in decisions that have to be made (Field Museum of Natural History 1976).

The opening of the Delaware Museum of Natural History in 1972 was also arranged with a focus on ecology and environmental awareness. An earlier but similar initiative can be observed in the opening of the Oakland Museum of California in 1969. This institution, formed for the residents of Oakland to exhibit natural science, history and art, reflected the concerns of the period. Divided into three floors, the first floor, which welcomed visitors to the collections, was named the Hall of California Ecology and displayed the natural history of the state. The exhibits, displays and dioramas were arranged over 38,000 square feet and focused on demonstrating the interrelationships of plants and animals across the varied environmental zones of California. These different areas were linked through the unique aspect of the Hall of California Ecology, its walkway, which offered a trail through the natural history of the state. Situating its visitors within a natural surrounding and highlighting the impact of humanity on the delicately balanced ecosystem within California, the Oakland Museum placed visitors into the same moral quandary as the Field Museum's exhibition: how do we act towards the natural world? The same issues were being confronted by museums across the rest of the world. During the 1960s and 1970s, the Naturhistoriska Riksmuseet (The Swedish Museum of Natural History) in Stockholm used this focus on contemporary environmental issues as part of programme of community engagement (Engström and Johnels 1973).

By the 1970s and 1980s, sites of natural heritage were also becoming the focus of critique themselves for the portrayal of stereotyped and prejudicial ethnic and gendered identities within displays, arrangements, exhibitions and information panels. The assessment of natural history as a confining discourse that disenfranchised and disempowered communities critiqued the role of institutions which had established themselves as objective purveyors of public knowledge. Within the United States, the issue of racial identification within museums, galleries and archives became a point of conjecture as museums were observed to be the last bastions of unconstructed racism. This was not the case in all circumstances; indeed, some natural history museums had been the site of counter-narratives which challenged institutional and implicit racist presumptions. For example, the Los Angeles County Museum of Natural History held an exhibition entitled 'Black Heritage' in 1969 which sought to emphasise how African Americans had been prominent in transforming the natural and physical landscape of California (Bellous 1969). Nevertheless, the charge of racism was increasingly levelled against natural history museums by African American groups for their arrangement of displays, collection policies and exhibition programmes by the early 1970s (see Anon 1974: 8–9). The American Museum of Natural History was the site of protest in 1974 as members of the organisation Friends of African-American

Studies Committee labelled the display of natural and cultural exhibits from Africa as 'racist'. This group, who counted the Pan-Africanists Dr John Henrik Clarke (1915–1998) and Dr Yosef ben-Jochannan (1918–2015) amongst their membership, sought to transform this section of the museum as a place of cultural awakening for African Americans (see Peter Bailey 1975: 55). Whilst this provided an alternative mode of practice for the institutions, it continued the modern use of natural heritage as a means of defining the present and guiding the future.

From these debates, the function of sites of natural heritage as locales of transformation, where humanity could be altered and shaped as the changes in the natural environment were regarded, was further affirmed. To address the problems of the modern world, natural history could be used to demonstrate alternative values and perspectives. This was observed with the American Museum of Natural History which was the centre of further controversy during its highly successful, major exhibition programme 'Ancestors: Four Million Years of Humanity', which ran from April to September 1984. The cause of this dissatisfaction was the invitation extended to scholars as well as the use of fossils and specimens from museums and institutions which were all from apartheid-era South Africa. The objective of the exhibition was to bring together the scant fossil evidence available for hominid development to create a wider narrative about humanity (American Museum of Natural History 1984). In response to the exhibition, members of the City Council of New York voted to remove municipal funding to the museum. The Reverend Wendell Foster, a Democratic City Councilman from the Bronx, argued that the museum should be forced to remove any association with the racist policy of segregation conducted in South Africa (see Freedman 1984).

The New York Mayor, Ed Koch (1924–2013), was called upon to defend the museum's position and appeal for engagement and dialogue rather than confrontation. Intriguingly, it was to the common heritage of nature that this entreaty was made (see Freedman 1984). Whilst the bid to remove the museum's funding was unsuccessful, the campaign did exert pressure on the museum to be more overt in its conception of how the exhibition would make an impact on issues of equality and racism within the contemporary world. President of the Museum Thomas D. Nicholson (1922–1991) also stated that the exhibits of early hominids represented progressive values rather than implicitly supporting a repressive regime:

> The strongest statement the museum could make in opposition to racism or to any form of human injustice based on perceived differences between humans was in 'Ancestors' itself. The irrationality of racism and of prejudice based on racial differences was implicit in the exhibition. All the fossils and all the participants gave evidence in their materials and in their statements of the common ancestry of all humans (Nicholson 1984: 8).

This concept of contemporary relevance of natural history for environmental, political and cultural objectives altered the relationships between wider society and natural history museums. This issue was also exposed with the concerns expressed from the 1970s onwards that sites representing natural history also perpetuated gender norms and participated in supporting patriarchal structures of power. Between the early 1960s and the early 1970s, the 'Biology of Man Hall' was completed at the American Museum of Natural History which displayed the evolutionary adaptations of humans as mammals. Exhibits presented 'man as a species', focusing upon 'his relationships to other members of the animal kingdom', whilst also examining 'his evolutionary history' and 'his functions as an organism' (Anon 1972: 122). Whilst this scientific assessment of natural history was regarded as neutral, it appeared to some observers as a reaffirmation of modern gendered identities. In these galleries, the figure of 'Man' was the awarded the role of inno-vator, whilst the figure of 'Woman' was regarded with reference to rearing children. This particular mode of representing 'natural history' within a template of 'modern' understanding of sex and gender was also apparent in the 'Triumph of Man' exhibition at the New York World's Fair in 1964. The emphasis within this set of dioramas was man's conquest of nature and inexorable rise to greatness. Sponsored by the Travelers, the insurance company, the scenes replicated mid-century gender values onto every other era to demonstrate mankind's struggle over adversity:

> Out of a world of strange under-water cells, fish, primitive vertebrates that lasted for eons, came a gigantic breakthrough! A new creature appeared in this grisly stifling world. He is naked, groping, bent and ape-like. He is man – the most interesting, exciting, promising creature that ever walked the earth (The Triumph of Man 1964).

The dynamic presentation of this history of biological and cultural evolution obscured the rationalisation of contemporary gendered identities, skills and work as an inherent aspect of humanity. This construction of natural heritage as a vehicle for modern social ideals was clearly defined in the work of the Natural History Museum in London during the 1970s, as the new style of museological display was countered by critics who asserted that, despite the innovations, the same, traditional narrative of patriarchal hegemony was present. In 1977, the Natural History Museum launched a fully interactive display entitled 'Human Biology – an exhibi-tion of ourselves' (Anon 1977). This was the first part of an innovative development within curatorial practice that transformed the museum's mode of engagement with visitors over the next decade (Perks 2015). Known as the 'New Exhibition Scheme', this mode of representation for the 'Human Biology' display included films, slides and models which were all arranged within an immersive layout allow-ing visitors to move between spaces as part of their 'experiential learning' (Miles Head and Tout 1978). This arrangement was designed to evoke questions about human nature and the way in which humanity has survived in the past, present and future. Despite the accessible style of exhibition, the display attracted criticism from feminist and equality campaigners for its perceived disdain of women's roles (Sykes 1977: 45)

This particular critique was also applied to another exhibition at the Natural History Museum which focused upon human development. Entitled 'Man's Place in Evolution', it opened to the public in 1980 as part of the institution's new type of programming focused on education and engagement (Gosling 1980). The exhibition was centred on the themes of man as an animal, man's fossil relatives, man's living relatives, man the toolmaker, and modern man (Man's Place in Evolution 1980). Whilst the introductory panel of the display stated that the use of the term 'man' was meant as a synonym for 'people' or 'humans', this 'linguistic sexism' was assessed as a minor issue in comparison with the overt gender bias which was regarded as pervasive across the exhibition (Brackx 1980: 16).

In this manner, natural history is regarded as both the point of oppression but also the means of resistance. By the 1980s, sites of natural heritage across Europe and North America were central to schemes of education and public knowledge but also to campaigns for equality, social justice and sustainability. What is noticeable about these responses is the manner in which 'natural history' is regarded as a means by which change could be effected, as through this particular form of heritage society could redefine itself. This perspective could be allied to issues of gender, race, colonialism or environmentalism as 'natural history' has been asserted as a point of asserting progressive, inclusive values. This can be noted in the work of the Australian museums which have used natural history within exhibitions and displays since the 1970s as a means to address the legacy of colonialism in the inequality, underrepresentation and racism within wider Australian society (see Museums in Australia 1975). Whereas once the designation of 'natural history' delineated groups of people from one another, it was now provided as a means of developing a shared sense of heritage within the nation (see Edwards and Stewart 1980). Indeed, this notion formed the basis of the 1975 Australian Heritage Commission Act:

> For the purposes of this Act, the national estate consists of those places, being components of the natural environment of Australia or the cultural environment of Australia, that have aesthetic, historic, scientific or social significance or other special value for future generations as well as for the present community (Australian Heritage Commission Act 1975).

In an era of domestic political instability during the 1970s, natural heritage offered a point through which Australian society could address issues of tension and concern for the 'present community'. This marked a distinct departure from nineteenth century museums in Australia which placed Aboriginal Australians as part of 'natural history' as a means of oppression and to delineate groups of people from one another to support a racist social system (Lampert 1986). Within the new wave of exhibitions, a relationship to natural history was used as a means of generating greater awareness and respect for Aboriginal Australian culture as well as developing Australian national identity. At the Australian Museum, the Aboriginal Gallery was renovated and opened in 1973 but it was successively developed over the course of the next decade to reflect these shifting trends. In 1986, the 'Aboriginal Australia Gallery' was unveiled as a major advance in curatorship and consultation as communities of Aboriginal Australians were engaged with the display of objects and the formation of the exhibition narrative. The gallery was, therefore, forwarded as a means to counter the dismissive and derogatory perception of Aboriginal Australian culture that emerged within the colonial era and persisted in some sections of society. It is through the natural world that this past, present and future of Australia were negotiated. Indeed, the exhibition opened with a scene of the Australian outback and a depiction of the creator deity, the Rainbow Serpent, detailing the natural landscape of the nation and the relationship with the people who inhabit it (Anon 1985: 27–28).

This theme was explored further with the 1988 exhibition at the Australian Museum, 'Dreamtime to Dust: Australia's Fragile Environment'. In this major gallery piece, the natural history of Australia was explored across the millennia from 200,000 years ago to the present (Osborne 1988). The exhibition consisted of three dioramas, each replicating a natural scene from different eras of Australia's history. These walk-through arenas enable visitors to place themselves amongst the environmental conditions of the past. Whilst the earliest of these displayed extinct animals presiding over a primordial setting, the environmental conditions represented in the subsequent dioramas demonstrate the increasing impact that humans have had on the natural world. The relationship between the environment and the Aboriginal Australian populations of 50,000 to 60,000 BCE is highlighted as one of negotiation before eventual equilibrium. The display panels and dioramas therefore evidence a settled and respectful attitude towards nature within Australia's past which is highlighted as disintegrating with the arrival of European settlers. These dioramas lead to a point of reflection for visitors as they were asked to contemplate attitudes towards environmental issues and their role in caring for and protecting Australia's natural heritage. The delicate balance of the contemporary Australian ecosystem was presented as a means of orientating visitors towards a recognition of both Aboriginal Australian culture and the need to reform society through a greater regard of natural history. The harmony that was formed within Aboriginal Australian communities was offered as a model through which Australia could be reformed to ensure its development and sustainability. In that confrontation with modern concerns of inclusion and sustainability, it is natural history which is evoked to provide a point of community, identity and consolation.

The transformation of Australian heritage and museum service from 1975 onwards also paved the way for the formation of a national institution. The Museum of Australia was established by an Act of Parliament in 1980 as a means to display the 'history of the interaction of man with the Australian natural environment' (National Museum of Australia Act 1980). Whilst construction was not initiated until the 1990s, the earliest conception of the purpose of the museum drew upon natural history to establish a common identity amongst visitors. Indeed, the three major themes of the museum offered the natural world as a bridging point where past and present could meet and the future be redefined. The museum was detailed with addressing 'Aboriginal history', 'non-Aboriginal history' and the 'interaction between man and his environment'. It is the latter which was forwarded to encourage a sense of a multiculturalism and acceptance within Australian society as it gave a common arena through which connections could be formed. Natural heritage was both a means of understanding the past and an issue of concern and solidarity in the present (see National Museum of Australia 2001). Indeed, it is this mode of representation within sites of natural history that has dominated museums, parks, gardens and heritage sites. As the uncertainty of the future is regarded, the natural heritage of nations and the entire globe is used as a point of reflection.

Over the past four decades, a common theme regarding environmental change and human responsibility within natural heritage has been present. From the marking of the first Earth Day in 1970 to displays about pollution, erosion, deforestation, acid rain, climatic deterioration and mass extinction events, the perspective offered by natural history has been used to confront the fears and anxieties of the modern world. Indeed, in such circumstances it is perhaps not surprising that, with the rise in concern that environmental catastrophe, as a consequence of industrial growth, could limit humanity's progress or even ensure its destruction, the interest in previous extinction events has increased (see Mitchell 1998). Dinosaur exhibitions and displays experienced a boom in popularity from the 1970s as comparisons between the destruction of one group of animals and ourselves were drawn. This could be observed in the Natural History Museum, London, with their exhibition 'Dinosaurs and their Living Relatives' in 1979. Whilst the display explored the taxonomic relations between contemporary species and the dinosaurs, it is the nature of survival and adaptation with climatic change that underlined this exhibit (Anon 1979). Part of this process has been the development of an empathetic vision of the dinosaurs, where the extinct creatures are rendered 'real' or 'familiar', as the animals whose fate we might share are shown to be comparable to ourselves.

This relationship with this specific part of natural history was explored with the 1986 'Dinosaurs, Past and Present' exhibition organised by the Los Angeles County Museum of Natural History. This display of models, paintings and sculpture from across the world detailed the history of humanity's understanding and perception of dinosaurs over the past 200 years. The alteration from terrifying lizards and ponderous dim-witted beasts to victims of an environmental apocalypse stems from both developments within scientific knowledge but also the relationship that is formed with this natural history (see Czerkas and Olson 1986). A similar exhibition regarding the place of fossils in mythology was also explored in 1988 within the Naturmuseum Senckenberg in Frankurt (Ziegler 1988: 18). The 1991 animatronic exhibition at the Natural History Museum, London, and then at various international venues, 'Dinosaurs Alive!' provided an opportunity for visitors to encounter a series of moving models of various dinosaurs as a means of considering the biology and the behaviour of these creatures. To further evidence this process, in 1995 the Hall of Dinosaurs within the American Museum of Natural History, New York, was remodelled for the purposes of public information. The express purpose of the new modes of engagement with dinosaurs which detailed the lifecycles and habits of the creatures was to provide clarity on a topic which, whilst well known, was regarded as poorly understood (Norrell, Gaffney and Dingus 1995). At a point when humanity's survival as a species is threatened, there is an apparent solace in acquainting and familiarising society with the previous dominant animal on Earth.

Similarly, this perspective was also evident in exhibitions and displays that highlighted the extinction of species across the world as a consequence of human activity. In 1988, the Smithsonian Institution developed the travelling exhibition 'Tropical Rainforests: A Disappearing Treasure', which toured across the United States within natural history museums and other public institutions until the early 1990s (Anon 1988). This extensive display, which included dioramas, videos and interactive displays, was designed to highlight the destruction of natural habitats, the threat of extinction for species as a result of the expansion of logging and agriculture and the impact of western lifestyles on the rainforest. By demonstrating the immediacy of the problem within the context of natural history, the relationship between past, present and future was brought to attention for visitors (Avent 1990).

The same issue was also explored earlier within the Cleveland Museum of Natural History's exhibition 'Confiscated!' in 1980. This display of items of modern humanity's consumer obsessed society which were all obtained from the natural world, from furs, hides and rare woods, demonstrated the exploitation of resources. The pieces within the exhibition had been seized by customs officials within the United States, as the exhibition exposed how 'commodity' status applied to the natural world fuels the destruction of habitats. The display was part the museum's programme of raising the profile of this issue and it also travelled to several other prominent natural history museums to attract the wider public's attention. The human impact on the environment became a constant part of the public engagement programmes within sites of natural heritage throughout the 1990s.

In advance of the 1992 Earth Summit in Rio de Janeiro, the American Museum of Natural History launched a joint initiative with the Environmental Defense Fund to create an exhibition and touring display entitled 'Global Warming: Understanding the Forecast' (Revkin 1992). This hands-on, interactive and educational piece detailed the way in which greenhouse gases had been created by industrialisation and urbanisation. Significantly, visitors to the museum were instructed in ways they could reduce their environmental impact. The institution's context thereby accentuated the values of the exhibition as it lent credence and authority to what was at this point still a highly debatable issue. Whilst the topic of greenhouse gases was defined, the museum was able to place the developments of climate change within the surroundings of an established narrative of humanity's depletion of resources. These concerns were also explored by the Natural History Museum, London, which launched a new Ecology Gallery in 1991 that explored the issue of global warming in relation to greenhouse gases, as well as the threat posed by the destruction of the rainforests and the depletion of the ozone layer (see Tudge 1991). Natural history becomes a means of consolation, to render the anxieties of the modern age into a comprehensible format. It is to a sense of natural heritage, therefore, that an appeal is made in order to direct attitudes and ideals. It is this sense of place, time and identity that sites of natural heritage are invested with that enables their elevation as places where the past is regarded as a means of addressing the future. It is this status that United States Secretary of State Madeline Albright referred to in her address on Earth Day, 21st April 1998, at the National Museum of Natural History Washington, DC:

> Thanks to the Museum of Natural History for sponsoring this event. From the Hope Diamond to the blue whale to the multi-legged attractions of your Insect Zoo, the wonders of the world are on display here; and I can think of no better place than this to recommit ourselves to our planet's environmental health (Albright 1998).

This speech, delivered a few months after the Kyoto Protocol, provided a rationale for the highly debated decision by the Clinton administration to endorse the commitment to reducing harmful greenhouse gases in the United States (Anon 1999a). It also conveys the power of sites of natural heritage to transform the present, as these museums, parks and gardens serve as locales of reflection and perspective. To regard contemporary concerns from the vantage of 'natural history' is to perceive the future as much as it is to understand the context of current issues.

This use of heritage has been used in other programmes of social reform beyond environmental concerns as sites have mobilised wider perceptions of natural history to contextualise current fears, anxieties and threats and direct society towards the future. Indeed, it is through this mode of engagement that pressing global concerns have been addressed. In this manner, natural history museums have served as an important platform in raising issues of world health. With public concern and government programmes initiated after the identification and emergence of HIV/Aids as a social health issue during the 1980s, sites of natural heritage became locales where the shock and trauma of this crisis was addressed. In 1988, the American Museum of Natural History was host to the exhibition 'In Time of Plague: Five Centuries of Infectious Disease in the Visual Arts'. This arrangement of artistic responses to the history of viruses and their impact upon the human condition provided a means of rationalising the panic and addressing the prejudice that ensued with the public awareness of HIV/Aids (Fox and Karp 1988). This point was developed further with the 1999 exhibition at the American Museum of Natural History, 'Epidemic! The World of Infectious Disease'. Within this display, interactive panels, videos and models were used to illustrate the history and contemporary concerns within the study of epidemiology. A range of infectious diseases, including tuberculosis, malaria and HIV/Aids, were presented to visitors as part of a web of life on Earth that have evolved and adapted. Within the exhibition, visitors could track the development, spread and modes of control for these diseases (DeSalle 1999). The value of this 'Epidemic!' exhibition was that its displays and models became part of a travelling exposition which was disseminated across the wider United States and funded by a variety of sponsorship schemes. This engaged local communities with specific public health concerns in specific parts of the nation which through the context of 'natural history' could be placed within a navigable and usable context. For example, during the spring and summer of 2001, the exhibition was located in the San Diego Natural History Museum. The exhibition acquired particular relevance in a city which had high rates of tuberculosis infection and poor access to health services.

In such circumstances, enabling present-day concerns to be placed within the context of 'natural heritage' provided a degree of familiarity and comprehension for local residents with regard to health concerns which may otherwise go unrecognised. The alteration of context and perspective which is developed with the sites of 'natural heritage' has been used for other forms of socially progressive objectives as well beyond the remit of public health and into the realm of cultural values. As such, sites of natural heritage have been used as a means of validating, defining and respecting identities (see Ebeling 2002). Museum exhibitions and displays have been key in this process as exposing the 'natural history' of groups who have suffered persecution or underrepresentation is forwarded as a dissident narrative. This was noted in the 2006 display at the Naturhistorisk museum in Oslo which was entitled 'Mot naturens orden: En utstilling om homoseksualitet i dyreriket' (Against nature? An exhibition on animal homosexuality). Through a series of case studies from the natural world, the museum detailed the existence of same-sex relationships within the animal kingdom in a direct attempt to confront prejudice and homophobia (Naturhistorisk museum 2006). In 2011, the Natural History Museum, London, produced an exhibition named 'Sexual Nature' and a touring display which was also designed to highlight the significance of difference within sexual behaviour across the species. This exhibition drew from the extensive collections of models, artwork, illustrations, video and taxidermy in the museum to show how animals mate. The culmination of the display was a section on human sexuality and the inhibitions, drives, attractions and practices that constitute the variety of sexual practice (Natural History Museum 2011). The recognition of difference within the natural world, both in the past and the present, is forwarded as justification of difference within contemporary society. In effect, the rationalisation of the modern world, the consolation of anxieties and the future direction of society are all realised within natural heritage. It is in this seemingly most stable aspect of our environment that we observe and define the most change.

Conclusion

From the curiosity cabinets, museums, gardens and parks, sites of natural heritage have been formed and defined over the course of the modern era. Indeed, it is within these places that the modern world has been shaped. In its earliest guise, as locales of knowledge and distinction where the act of collection imposes a sense of order and rationality on the wider world, natural heritage has been used to justify and endorse wider social and political relationships. As locations of beauty and wonder, where the act of observation exercises aesthetic and scientific judgement, the natural world has served as a medium through which society can be reformed. As the subject of exhibitions and displays, natural history has been cast as a point of reflection where individuals, communities and nations are orientated towards the future. What is evident from the survey of the form and function of natural heritage over the course of the last few centuries is that it is through this particular element that western society has defined itself as 'modern'. This notion of 'modernity' has been characterised by addressing current anxieties and preparing for how society could be organised in the future. This has enabled repressive and authoritarian policies to be pursued but it has also provided a means of advocating progressive reform. Whilst the political objectives might seek to mobilise natural heritage in a variety of ways, what is distinctive is its function. Drawing upon the context of this heritage, museums over the last century have formed and used natural heritage as a conveyor of meaning and as a place of witnessing. Whether this act of witnessing observes the awesome power of nature, the organisation of the natural world or the comparisons and lessons from nature that can be drawn for our own era, this is a mode of engagement with heritage that is utilised to direct power, place, authority and identity. Therefore, this association with the past and its legacy has been fundamental to understanding both the present and what is yet to come. In an era characterised as anxious or risk-adverse, where the threats posed by environmental disaster or ecological degradation are widely acknowledged, the perspective enabled by natural heritage will become critical in how we define ourselves, our relationship to each other and our relationship with the wider world.

3   Museums and natural history

Introduction

The array of institutions that display natural heritage appear to be strangely neglected as a subject of a critical heritage studies. Such a situation belies their significance to contemporary society. The engagement with this aspect of the past reveals attitudes and ideals regarding the present day and the future. Visitors to these sites are witnesses to vast timescales of history, complex processes of adaptation and development and intimate narratives of extinction, survival and change. The museums and exhibition centres that represent natural history provide a global, national and local perspective on these points of concern to visitors. As such, these locales are key points in asserting and altering social, cultural, political and economic norms. Through the display cases, signage, exhibition cabinets, engagement portals and media stands, there is great diversity across global institutions in the manner that this heritage is communicated and interpreted. However, what is consistent in these sites is the way in which they orientate this heritage within particular 'tenses' for the wider public to observe, acknowledge and bear testimony to. It is this function of contemporary museums, visitor centres and exhibitions of natural heritage which will be the focus of this chapter as the uses of this history are considered. Collections of fossils, geological specimens, ancient and modern flora and fauna as well as recreated environmental scenes and reconstructed settings are all featured as part of this assessment. Drawing upon case studies from across western society, using international, regional and local institutions, this study will detail how we guide current society and shape our expectations of what will happen in times ahead through this particular heritage.

Values and ideals: museums of natural heritage

The formation of museums detailing the natural history of nation states was a phenomenon of the nineteenth century which has left a legacy within the capital cities and major metropolitan areas of Europe and North America. These sites, which once affirmed the progress and status of the nation both culturally and scientifically, still retain the powerful, prestigious status of their past but their role in the present is framed by a stated desire to shape the attitudes and opinions of their visitors. Over the past twenty years, national museums of natural history have moved towards socially progressive goals as a means of addressing current environmental concerns. A point of connection between these types of institutions in the western world is their commitment to becoming 'sites of conscience'. This term emerged with the development of human rights museums which mark oppression, genocide and trauma. Equating major natural history museums with this association asserts the understanding of these sites as transformative spaces where society can 'learn' from past mistakes and create a better world. Indeed, the Natural History Museum in London states that its mission is to challenge the way society sees itself, to stimulate debate and to raise scientific understanding as a means of ensuring the future (Natural History Museum 2016a).

Merging scientific objectives, humanitarian missions and moral imperatives, the museum sets itself as a radical, all-encompassing venue where visitors can engage with their immediate environmental pressures whilst observing their place within a wider cosmic narrative. Significantly, the objectives of the institution are formed from a representation of the past with a specific regard for the way in which individuals and communities inhabit the planet today and how forthcoming generations will manage the environment in the years to come. Essentially, it fulfils that ambition of modernity as defined by Giddens (1990), which is to preserve and extend the objectives of contemporary society: to 'colonise' the future. This objective is certainly not unusual within the national institutions in Europe and North America. The Muséum National d'Histoire Naturelle in Paris asserts the same claim in its mission statement, bridging the millennia and making what is at times a distant past exceedingly relevant:

> The Museum has a wide field of engagement, where the study of the past illuminates the future and which combines the sciences of nature and of man. It carries out these activities through five missions, performed by various professions and disciplines, enabling it to examine this subject in all its dimensions (Muséum National d'Histoire Naturelle 2016).

Intriguingly, both institutions regard their ability to guide visitors through the past, inform the present and shape the future through their place as centres of knowledge production. The notion of authority is asserted as the rationale for these institutions as the museums are presented as the vanguard of a scientific mission (Bennett 1995: 126). The five missions stated by the Muséum National d'Histoire Naturelle reflect this positioning of the institution: 'fundamental and applied research; preservation and collection development; teaching; dissemination of knowledge; expertise' (Muséum National d'Histoire Naturelle 2016). The symbol of scientific and intellectual authority, so significant to the institution in its initial formation, is still emphasised here as a rationale for the museum's social role of 'illuminating' humanity's way ahead. The emphasis on the expertise and authorial voice of the museum is key here as it establishes the right to inform and shape societal issues. The Museum für Naturkunde in Berlin establishes the existence of this legitimating discourse within natural history institutions with its future-orientated, authoritative mission statement that combines moral concern with intellectual rigour:

> We study life and planet Earth, maintaining a dialogue with people. Our mission, our vision, our strategy and our structure make our Museum an excellent research museum .... we have become an innovative communication centre that helps shape the scientific and social dialogue about the future of our earth – worldwide (Museum für Naturkunde 2016a).

The definition of these sites as instructional and as spaces of reform for society creates a manner of framing major natural history museums as arbiters of taste, distinction and value. In effect, these institutions have developed from their origins as bastions of nineteenth century nationalist and imperial identity through affirming and shaping visitors' sense of identity only to return to the same function within contemporary Europe and North America at the outset of the twenty-first century. Significantly, this elevation of social mores and the advancement of moral and scientific concern is undertaken without recourse to the nationalist framework upon which these institutions were formed. Indeed, what emerges at these natural history museums is concern for a transnational identity focusing upon global humanitarian concerns. As such, visitors to these sites become moral and political witnesses. This is the key for these institutions to act as sites of conscience. Elevating collections and displays beyond the confines of the traditional interpretations associated with these materials enables museums to serve as a key feature in the modern world. A global citizenship through this heritage is offered to visitors as they are asked to consider the museums' work within a wider context (after Hein 2000). The Museo Nacional de Ciencias Naturales in Madrid is characteristic of this orientation, as it places its work at the forefront of academic research but advocates a role of public awareness beyond the municipal and national concerns:

> Our challenge today is to convey the knowledge generated by our researchers to society and for this we have highly qualified professionals dedicated to both scientific collections that houses the museum and the exhibits that allow us to explain scientific discoveries to the public that visit us (Museo Nacional de Ciencias Naturales 2016a).

The significant aspect of this global heritage citizenship within natural history museums is the delineation and realisation of common traits that define this collective identity. An assessment of the missions and objectives of national institutions within contemporary Europe reveals an idealised vision of what a visitor to natural history museums would represent. A scientifically aware, future orientated individual who is actively engaged in environmental issues is the model global citizen for these heritage sites. As places of 'conscience', institutions provide a series of instructions to frame the wider understanding of natural history as a tool for use in the present. This is demonstrated within the work of an organisation such as the Statens Naturhistoriske Museum in Copenhagen, Denmark, which forwards its collections, exhibitions and displays as a vital means of engagement with the public to understand the value of science and the significance of sustainability (Statens Naturhistoriske Museum 2016). Similarly, the Naturhistorisches Museum in Vienna (Wien) emphasises its role as a public institution to guide contemporary society through the way in which our world has been shaped by geological and climatological processes whilst detailing the significance of this work in the present (Naturhistorisches Museum Wien 2016a). These objectives can be located within the venerable institutions of the imperial era across European cities, where a new language of internationalism is found within the focus on research and public knowledge. The colonisation of the future is evident here as the particular type and form of relationship that society should have with the world is defined at these sites. The Muséum des Sciences Naturelles de Belgique/Koninklijk Belgisch Instituut voor Natuurwetenschappen (MSNB/KBIN) in Brussels, Belgium, is noticeable for this objective, expressed clearly in its mission statement:

> The museum plays a leading role in the promotion and dissemination of scientific culture, both within and beyond its walls, notably through travelling exhibitions and events. We are pursuing our ambitious efforts to gradually renovate the premises, making the museum more welcoming and meeting and exceeding our visitors' expectations. We also seek to promote a more respectful approach to nature (MSNB/KBIN 2016).

Such a consensus amongst museums in Europe would indicate an implicit or explicit drive to conformity within this heritage. The socially progressive agendas of institutions of natural history are dependent upon possessing a shared vision within the museum (Hein 2000). Visitors are encouraged to consider the museums as locales where a sense of self, purpose and direction is created (after Sandell 2016). Indeed, the museums of natural history are imbued with significance as a place of value and worth in itself as it is within this space that a reflective, global identity is formed (Golding 2009: 16). The Yale Peabody Museum of Natural History in New Haven, Connecticut, is part of this movement towards a sense of fellowship in the face of increasing threats to the environment. Therefore, this nineteenth century institution has taken on a twenty-first century mission through the formation of a universal character of custodianship for the past, present and future:

> Fundamental to this mission is stewardship of the Museum's rich collections, which provide a remarkable record of the history of the earth, its life, and its cultures. Conservation, augmentation and use of these collections become increasingly urgent as modern threats to the diversity of life and culture continue to intensify (Yale Peabody Museum of Natural History, 2016a).

The central component of this sense of self appears to be an 'environmental consciousness', or a moral witnessing, where visitors are required to develop an awareness of their place within the wider ecosystem (see Hussey and Thompson 2000).

The repetition and rehearsal of the role of institutions in education and research appears to create a degree of homogeny within these sites. Whilst visitors may be drawn from all over the world, with varying cultural backgrounds and political beliefs of a variety of hues, they are still required to respond to these institutions in the same manner. For the purposes of environmental protection and sustainability, a disciplinary agenda is inevitably introduced to model wider society (after Foucault 1979). Whilst discipline of taste, distinction and ethics have emerged within museum studies in recent years, what is defined here is an institutional discipline that cultivates a sense of being and place. Institutional discipline reinforces the authority of the museum and the necessity of visitors acquiring a suitable awareness of their environment. Within these places, individuals are obliged to undergo a transformation which will enable them to sustain the present and protect the future. For example, the Naturhistoriska riksmuseet (2016) in Stockholm, Sweden, describes its mission to forward and improve knowledge of the environment in order to inspire people with the need to care for the planet by creating a space for experts and the public to exchange ideas. Therefore, these institutions form a site of moral witnessing where shared attitudes and ideals are formed and reiterated.

This follows from Hein's (2000) definition of 'institutional morality', which asserts the ability of museums to possess moral agency through their orientation, outlook and outreach. In this study of institutions, Hein also highlighted the tendency of modern museums to reassert the same ideals and values across national boundaries and disciplinary agendas as an inevitable consequence of globalisation. The insistence on singular visions within the museum was regarded by Hein as ultimately self-destructive as it was only through the recognition of a multi-vocal engagement that commonly-held principles and morals were discussed and developed (Hein 2000). The imposition of the 'environmental consciousness' across contemporary European and North American natural history museums ensures the appearance of a stark contradiction, where sites of Enlightenment culture attempt to reorder Enlightenment ideals through Enlightenment philosophies (after Lord 2006). The revolution in the use of museums within western culture that this represents is astounding but there is a danger of reasserting traditional museological relationships or failing to escape the 'genealogy' of institutions (after Bennett 1995). As the 'environmental consciousness' defines a global citizenship that is formed through scientific and moral awareness, the representation of natural history serves as a civilising mission just as it once did in the nineteenth century. A set of behavioural norms are present here which are expected to be fulfilled. The extent of this approach within European and North American museums is overwhelming as across the continent this is a commonly-held assumption within institutions. Natural history museums continue to fulfil the Enlightenment project of rationalising society. This is observed in the value statement issued by the Eesti Loodusmuuseumi (2016) in Tallinn, Estonia which emphasises its 'world class exhibitions and collections' as part of a 'well-recognized centre that shapes attitudes into ones that value the natural environment'.

This collective identity is formed from the construction of a mode of witnessing at these sites through a moral frame (after Rüsen 1990). Institutions shape the role and function of the witness through their orientation as collectors of the past and as authorities on the present whilst asserting their didactic role in educating and informing visitors. The moral frame in which visitors are brought into the museum is the focus on the contemporary relevance of the collections. Objects and items are heralded with a quality of prescience which can serve to guide the present. It is this immediacy and neoteric aspect to subjects and items which may be several millennia in age that is common throughout natural history institutions. By making their collections present, these museums claim their place in the wider intellectual and moral landscape as progressive and cultivating centres for society. This is reflected in the Nederlands Centrum voor Biodiversiteit Naturalis which emerged after 2011 from the restructured Zoölogisch Museum Amsterdam which was formed in the nineteenth century. In this new museum, an old museological discourse is asserted for the purposes of modern society:

> At Naturalis Biodiversity we want to describe, understand and explain. For the welfare of the people and the survival of our life on this planet.
> (Nederlands Centrum voor Biodiversiteit Naturalis 2016)

This function of the museum requires a mode of moral witnessing that is orientated towards immediacy. Indeed, institutions assert the requirement to alter our relationship with the wider environment as a matter of urgency. The natural world is presented as threatened and on the brink of irrevocable change or irreparable damage which can be addressed through engagement with this natural heritage. As institutions detail the importance of their collections, a focus on contemporary issues is palpable. The Canadian Museum of Nature in Ottawa typifies this emphasis on action that arises through modern institutions representing natural heritage:

> Knowing more about nature gives us the tools to make better decisions about resources. It provides the basis for new technologies and developments, and promotes a better understanding of how we affect, and are affected by, the natural world (Canadian Museum of Nature, 2015).

As such the political character of this mode of witnessing in terms of its immediacy is distinctly moralising in its conception. The increasing urgency with which museums place their mission statements, research and values reflects the certainty that climate change is altering the planet and humanity's ability to continue to survive in particular regions. However, this sense of speed and rapidity is also symptomatic of the conditions of late capitalism where the demands for action and reaction are almost instantaneous (see Noys 2014). It is this malaise that Žižek (2011) characterises as evident of western capitalist society's urge to correct its excesses but maintain its structure. In essence, this preserves the underlying socio-economic form whilst attempting to alleviate some of the consequences of that mode of production. As Žižek states, 'today, we do not know what we have to do, but we have to act now, because the consequences of inaction could be catastrophic'

(Žižek 2011: 480). In the rush to do 'something', to alter patterns of behaviour to act on new reports, we cease to understand how this situation initially occurred. Therefore, as institutions require a sense of immediacy, engagement and action through their mission statements and displays, the political witnessing of natural history as a current concern reflects more than the pressing need to address the issue of climate change. An act of political witnessing is apparent in this urge to change and alter which obscures the malleable effect of natural history and the potential for alternative uses of that heritage. This is not to diminish the progressive, environmental goals of these objectives, but rather to state that the process of reaching that objective can be reimagined in the context of a wider set of concerns regarding the structures of society.

As this moral frame constructs the role of a specific type of witness within natural history museums, the perspective of the social witness is formed through the elevation of communities within these institutions. Whilst a sense of global citizenship confers a moral witnessing within museums, a social witnessing focuses upon the national, regional and local bonds which are formed within sites representing natural history to the wider public. Within European and North American towns and cities, natural heritage museums demonstrate their role within communities at a variety of scales to encourage the formation of a sense of place at these locales. Acting as a social witness to this heritage acknowledges the connections to others through these institutions. Significantly, this work is undertaken as a means to form a social contract between museums and their visitors. Rather than a singular moral vision or a politically expedient solution to wider economic concerns, the action of the social witnesses within natural history museums builds relationships between institutions, objects and communities. Essentially, it is this mode of observation that enacts the 'parliament of things'; ensuring that human actors and non-human agents form mutually beneficial connections amongst each other (Latour 2005). The 1999 Council of Europe seminar organised to assess the management of natural resources made this assertion:

> In any structured society, of humans and often even animals, there is a concept of shared wealth or common heritage which must be preserved if the society in question is to survive or at least function properly (Kiss 1999: 5).

The creation of social witnesses to this heritage can be observed in the variety of ways institutions orientate their objectives towards their immediate audience. Such an approach could be regarded as somewhat of an extension of the ecological exhibits and designs that emerged within museums from the 1960s onwards. However, rather than enable the observation of such interconnected systems in action by visitors through displays, a number of natural history museums are specifically orientating themselves towards becoming sites for this engagement to take place. The recognition of a changing, ecological model brings with it a means through which wider society witnesses the relationships they possess with others. This can be seen with the vision statement of the Natural History Museum, Los Angeles County:

Human beings are connected – to each other, to communities, to other species, and to the Earth. As humans increasingly influence natural systems, it is critical that we understand these relationships. This understanding, in the context of the history of the Earth and its inhabitants, guides our approach to investigation and interpretation (Natural History Museum, Los Angeles County 2016a).

Rather than a didactic assertion of moral goals, the formation of the museum as a locale where such connections can be discovered, dismissed, created and imagined provides a forum to discuss and deliberate beyond the confines and definitions of the institutional structures that were defined during the Enlightenment (after Sandell and Nightingale 2012). Indeed, the categorisation that such terms imply ensures an inability to escape the well-established modes of moral and political observance within museums (Foucault 1969). The removal of a rigid form of what constitutes 'nature' or the role of humanity towards the 'environment' is exposed here as a wholly beneficial recognition which orders an alternative modernity (Latour 1993). It is this concept that structures the Latourian understanding of the environment:

> The concern that one can have for it appears with the disappearance of the environment as what is external to human behaviour; it is the externalized whole of precisely what one can neither expel to the outside as a discharge nor keep as a reserve (Latour 2004a: 241).

This would constitute a 'social understanding' of natural history museums. To serve as a social witness at these sites would entail removal from the vaunted position of scientific research and the authority of the museum and to make the objects and issues of natural history democratic in form. What were regarded as universal truths become part of a process of dialogue (Latour 2004a). A number of museums have alluded to this process as a stated aim for their development within the last decade as natural history institutions respond to concerns about their role in addressing global environmental change (see Alberch 1993; Futter 1997). For example, the Natural History Museum of Utah, which was formed in 1969, describes some of its core values as processes of democratic engagement between communities and the wider environment:

- Demonstrate the myriad links connecting the past, present and future.
- Transcend scientific disciplines to reveal the networks inherent in nature.
- Serve as center for science literacy, acting as a bridge between the scientific community and the public.

(Natural History Museum of Utah 2016).

The effect of acting as a social witness to this history is to become engaged in the ways in which it alters both individual and wider societal perceptions. To be a witness in this manner is to participate in an exchange which is acted out within the museum. Indeed, the Natural History Museum of Utah specifically offers itself as a place for this dialogue to take place:

> The Museum provides opportunities for students and visitors to better under-stand their interactions with the natural world and provide a framework for individual and community decision-making (Natural History Museum of Utah 2016).

The description of the museum as a venue of democratic engagement and explor-ation enables the visitor as a social witness to observe and testify as to the signifi-cance of this account. A distinct departure can be noted in this process apart from the moral and political means of bearing witness. It is in this account of social witnessing that objectives and ideals may be set but they are regarded as the result of the relationships formed with the space of the museum, the collections and wider society. Intriguingly, these agendas are frequently presented within regional museums rather than national institutions. The framework offered within the latter, seemingly derived from the Enlightenment ideals of their foundation, could per-haps prevent the realisation of an alternative, social form of knowledge within these sites (see Kuhn 1962). This model of natural history research deters the rela-tionships that are formed between people and things by assuming a stable and solid field of study (Popper 1962). By acknowledging the constructed manner of this process as well as the dynamism of the subject matter itself, the social witness is able to make connections within the nexus through which museums orientate themselves for the present and the future (after Feyerabend 1975). This can be observed with the mission statement of the Cleveland Museum of Natural History:

> Explore our exhibits, meet our wild animals, take a hike, join a class – all while creating meaningful memories. . . . To inspire, through science and education, a passion for nature, the protection of natural diversity, the fostering of health and leadership to a sustainable future (Cleveland Museum of Natural History 2016a).

It is in this relationship with the museum collections that the visitor as witness can relate, observe and adapt as they testify to the significance of the institution to wider society (see Krishtalka and Humphrey 2000). Such approaches to the study of natural history are increasingly common within mission statements and value statements of institutions. As the environmental and economic concerns of the twenty-first century expose the strains and tensions of humanity's relationship to the wider world, institutions have sought to develop these formats to address current issues. This position is certainly not an anathema to the role of a global science programme which many natural history museums strive towards. Rather, it is a recognition of the wider social world in which this endeavour participates. Therefore, institutions which have a modern scientific research agenda, such as the Beaty Biodiversity Museum in Vancouver, Canada, can describe their role as providing a space where a sense of 'collective responsibility' is formed (Beaty

Biodiversity Museum 2015). Such positions are also incorporated into the global bodies which advise and guide institutions on their outlook and development. The International Council of Museums (ICOM) and the sub-committee detailed to oversee these specific types of institution, the International Committee for Museums and Collections of Natural History (NATHIST), agreed on a series of ethical guidelines in 2013 which specifically focused on this 'social' perspective:

- Build and store natural history collections;
- Conduct research and interpret the results;
- Support the process of science and biological conservation;
- Enhance public understanding and appreciation of the natural world;
- Collaborate with the public in deriving their own meaning from the natural heritage they encounter in the museum and in nature.

(ICOM 2013)

It is through this relationship, which sees the development of meaning and values, that the natural history museum is able to gain contemporary relevance and involvement in the future. Instead of asserting a model of development, a moral character or a political ecology, a place of orientation and democracy is provided. Increasingly, such approaches have been used by 'non-traditional' installations to challenge the operation of power and authority within museums. This has been a feature of The Natural History Museum; this initiative, a mobile, pop-up exhibition, which was a collaboration between artists and museum professionals that toured sites in the United States from 2014, offered social engagement and interaction as a central part of their work (The Natural History Museum 2016). In their displays, workshops and events, visitors are provided with reflective spaces rather than didactic ideals:

> The Natural History Museum's perspective on natural history differs by taking its orientation from two basic insights: nature is common and what is common is divided. We struggle over what is common. We fight to keep it common. The fact of this struggle alerts us to division, conflict, antagonism: nature has never been in balance. Nature doesn't just exist. It insists beyond the limits of the known. What we can't see and don't know impresses itself on how and what we see (The Natural History Museum 2016).

The specific concern of developing a sense of the 'common' reasserts the Latourian definition of a democratic society of things (Latour 2005). However, these concerns are not the preserve of those from a dissonant agenda or from the political periphery or the cultural avant-garde. Increasingly, this movement has become mainstream with institutions of varying sizes stating this approach within their objectives. Indeed, the mission statement of the National Museum of Natural History (SNMNH 2016a) in Washington, D.C., places its core value as '(u)nder-standing the natural world and our place in it' (Figure 3.1). The communal, collaborative and collective is promoted not as an alternative to the authoritative

Figure 3.1 Dinosaur bones, National Museum of Natural History, Washington, D.C.
Carol M. Highsmith Archive, Library of Congress, LC-HS503- 1569.

or as a challenge to the role of scientific research but as part of a recognition that alternative modes of dialogue exist and can be productive and progressive. It is through the social witness perspective that such means of engagement and access are exposed for greater assessment. Considering the way in which institutions form social witnesses to the past, present and future amongst their visitors provides a means of reconsidering the roles, values, identities and agendas of institutions.

The mission statements, objectives and 'visions' of museums relay the type of experience which is formed within the physical and intellectual space of these organisations. The moral witness is required to observe the pressing need for action within the contemporary world through the collections and displays of the museum. As such, called upon for a higher purpose, moral witnesses within these institutions inevitably require the affirmation of a singular truth regarding human activity and engagement with the environment. The justification of this approach is undeniably focused on generating awareness within society and improving humanity's relationship to nature but it succeeds in defining this relationship to a static phenomenon. If environmental change is occurring apace, then a certain degree of negotiation is required within this context as is a level of dynamism. This is reflected within the political witnessing present within museums, as visitors are informed by institutions to look upon collections to draw modern comparisons that inform, shape and accelerate action in the present. However, this regard for speed, response and immediacy also reflects a late capitalist ideology that ameliorates the symptoms but does not necessarily provide a panacea. It is within the act of social witnessing that acknowledges the multifaceted, active and fluid relationships that human society holds with the wider environment. Observing and testifying to the effect that individuals and communities are part of a wider dialogue where 'nature' is a fluctuating concept composed of aspirations and anxieties does not cast the world into uncertainty. It ensures that an enriched relationship with a sense of natural heritage is formed at a local and international level that can build associations with the past and provides a forum where current connections can be built, developed and altered to address the problems we face in common.

Witnessing the performance of natural heritage

The way in which objects and collections within natural history museums are presented to visitors provides alternative points of witnessing which shape the use of this heritage within contemporary society (after MacLeod 2005). Whether it is the organisation of exhibits, the layout of displays, the narratives provided for visitors or the modes of multimedia engagement within contemporary institutions, the witness perspective is evoked as museums seek to educate, entertain and engross their audiences. As witnesses, individuals and communities observe the representation of this history as a point of engagement with the natural world for the past, present and future. Despite the significance of this cultural heritage, the character, form and content of its depiction to visitors has been strangely neglected within wider scholarship as the dominance of a positivist framework within environmental studies and natural history has held sway (after Harré 1993). The critical examination of this area of investigation has only been undertaken relatively recently as increasing awareness of issues of climate change has brought the role of the natural history museum within the remit of heritage studies (see Dorfman 2011). Through the assessment of how witnesses to natural history are formed within these sites, the use of this heritage can be assessed as a construct which forms identities, legitimates power and forms a sense of place but also possesses the potential for dissonance and resistance (after Harrison 2013; Smith 2006). Through the display of flora and fauna, whether fossilised remains, reconstructed scenes or even animatronic displays, institutions place visitors in relation to this data as a means of bearing witness (Hein 2000). The effect of this witnessing orientates the social, political and moral relations that are formed within society. As such, it is within these spaces that human society can alter, adapt and change.

Perhaps the most dominant aspect of the national natural history museums in Europe and North America is the display of dinosaur collections to the public. Indeed, the fossilised remains, reconstructions or dioramic scenes of these extinct animals appear to have achieved a level of acceptance within society that their placing at the entry points of institutions is almost ubiquitous (see Chicone and Kissel 2014). Across a variety of national examples and case studies, the first object of significant interaction for visitors is evidence of life from the Cretaceous, Triassic or Jurassic eras. This is perhaps the legacy of the foundation of many of these institutions in the latter half of the nineteenth century when the popularity of palaeontology was at its height. It may also indicate the more recent revival of interest in dinosaurs which has been marked by scholars as emerging in the 1970s (Bakker 1975). Such fluctuating levels of public engagement with the subject stand in stark contrast to the apparent permanence which these remains and reconstructions are used to represent. With the mass extinction of the dinosaurs occurring over 65 million years ago, the representation of this history offers a demonstration of the tremendous depth of time that the institution represents. Significantly, the use of this immense chronological scale frequently asserts the authority of the museum in its position to detail wider aspects of natural heritage (after Macdonald 1998; 2002). The possession and display of such artefacts establishes the identity and character of the organisation. This is most clearly observed in the foyers and entrance halls of some of the world's foremost museums across North America and Europe.

For example, since the 1920s the grand hallway that marks the visitors' opening engagement with the American Museum of Natural History in New York has been dominated by a vivid, dioramic dinosaur display. Located in the Theodore Roosevelt Rotunda, the imagined scene is of a rearing *Barosaurus*, the long-necked herbivore of the upper Jurassic Period, defending a juvenile sauropod from the predatory instincts of an *Allosaurus*. The setting, composed of casts of fossilised bones, was remodelled in 2012 to allow visitors to walk through the opposing creatures as they are locked in a primeval battle for survival. The reflection of anthropomorphic desires upon the behaviours of these extinct creatures can be examined but what is significant to note is the immediacy and effect of time as a specific quality generated by such a display. Placed into a 'natural setting', these are artefacts that render millions of years of evolution into single momentary static phase. The act of interpretation in this manner is a demonstration of the institution's ability to inform, guide and educate through the presentation of time. Displays of dinosaurs and dinosaurian effects within the entrance halls of museums serves to communicate a sense of time perspective and frame an act of political witnessing. Visitors are presented with this reconciliation of chronological distance through the assertion of the authority of the museum. Time as a means of organisation and display requires observers to sequentially assess past and present into compartmentalised settings. When considering such vast timescales this might be regarded as a necessity, but it nevertheless reflects a particular 'time consciousness' identified with the emergence of the Enlightenment. In this conception, time is segmented into specific sections which enables the present to be understood as distinct and as the culmination of all the processes and events that have constituted the past. This was defined by Husserl (1991) as an awareness of 'past being':

> The past would be nothing for the consciousness belonging to the now if it were not represented in the now; and the now would not be now . . . if it did not stand before me in that consciousness as the limit of a past being. The past must be represented in this now as past . . . (Husserl 1991: 280).

The display of dinosaur remains or reconstructions within the entrance halls of museums and institutions affirms this notion of 'past being' as they place individuals as political witnesses to a crystallised point of another age which affirms those enlightenment ideals of progress. Cast in the reflection of 'time past' it is the progress of the present that is elevated. This process can be illustrated with the unveiling in May 2000, in the main hall of the Field Museum, Chicago, of what is regarded as the most complete specimen of *Tyrannosaurus Rex* ever recovered. Bought at auction in 1997 with assistance from McDonald's Corporation, Walt Disney World Resort and the University of California System, the display replaced the previous opening piece which was a *Brachiosaurus* that had dominated the entrance hall (Figure 3.2). The unveiling of the remains, replete with a plaster cast skull, was presented as a spectacle where the past could be observed by the present (see Miller 2000).

Whilst the event of the unveiling was made powerfully present, the object of this witnessing was placed into a distinctly 'other' era. The segmentation of time is accomplished in such a perspective as it emphasises the advancement of the modern age. Such acceleration from a 'past being' is also present in the character of 'Sue' which is attributed to the remains. The moniker is derived from the name of the palaeontologist who discovered the fossil in 1990. The anthropomorphisation of the fossil has been key in the museum's promotional and outreach work but it also serves to emphasise the vast time scales between the *Tyrannosaurus rex* and its human observers. The incongruity of naming the extinct animal, rather than assuming some emotional or social connection to the wider public, emphasises the place of the fossilised remains as chronologically 'distant'. The quality that such naming provides is a further assertion that the visitor witnesses an age long past upon which one can reflect and regard the advances of the current era. The Natural History Museum of Los Angeles County utilises the same technique in its Grand Foyer where the fossilised remains of a *Tyrannosaurus rex* and *Triceratops* are engaged in battle. Described as the 'duelling dinosaurs', the display locks an ancient spectacle in one specific moment in time for visitors to observe. In this act of witnessing, we are drawn to reflect upon this time as past and ourselves as present in an advanced state of evolutionary progression. In the

Figure 3.2 Exhibit hall at the Field Museum of Natural History in Grant Park, Chicago, Illinois. Carol M. Highsmith Archive, Library of Congress. LC-HS503- 6526

Denver Museum of Science and Nature, on their entrance into the foyer visitors are confronted with a *Tyrannosaurus rex* in what is often referred to as a 'dancing pose'. The naturalistic or 'realistic' display of these specimens is used to convey a sense of action for visitors to ensure that the fossilised bones do not appear to be museum exhibits but part of a living world. However, their use within the foyer, entrance halls and lobbies of institutions demarcates past and present and offers visitors an affirmation of the political witnessing of time past and present development.

Smaller scale museums in the United States with both a popularising and scientific mission also utilise this scheme with either complete fossil remains, partial specimens or statues and model replicas of dinosaurs. From the large institutions to the small local museums, the remains or reconstructions of dinosaurs are used to welcome visitors to the study of natural heritage. For example, a 12-metre skeleton of a *Tyrannosaurus rex* was placed in the stairwell of the University of California Museum of Palaeontology (UCMP), Berkeley, California, in September 1995. The specimen was displayed to visitors in a life-like 'active' pose in order for it to be 'biologically accurate', according to contemporary scientific research (UCMP 2015). The skeleton was also reinforced with steel to enable it to withstand earthquakes of varying scales of severity that strike in this region of the west coast of the United States. A team of students, scientists and film set designers were behind these aspects of the display. Protected against the contemporary environmental dangers, its presence secures for the museum a means of establishing its institutional relevance and the vast time scale of chronological distance between the visitor and the Jurassic Era creature. The Virginia Museum of Natural History (VMNH) utilises its main entrance, known as the Harvest Foundation of the Piedmont Hall of Ancient Life, to display a variety of fossilised remains. From the *Pteranodon*, a flying dinosaur from the Late Cretaceous Era, with a wing span of over 6 metres and arranged from the ceiling to appear as if it was diving onto those below, to a 140-million-year-old *Allosaurus*, the perspective of the political witness is reaffirmed in this arrangement (VMNH 2016). This is also demonstrated in the way in which the contemporary scientific analysis of these specimens is exhibited alongside the fossils:

> Visitors to The Hall of Ancient Life can also view VMNH scientists at work through windows looking into three labs. The Elster Foundation Vertebrate Paleontology Lab is where VMNH researchers prepare whales, dinosaurs and other vertebrate fossils, including over 700 dinosaur bones stored in the museum's collections from excavations in Wyoming (VMNH 2016).

Similar uses of dinosaur fossils within natural history museums can be found across the world. In the Oxford University Museum of Natural History, Oxford, Britain, visitors are immediately confronted with two cast skeletons of an *Iguanodon* and *Tyrannosaurus*. The looming presence of these large specimens is used to classify and order the wider species that are exhibited within the institution's collection (OUMNH 2016a). Indeed, the Museum für Naturkunde in Berlin, Germany, displays one of the largest dinosaur specimens in the world, a *Brachiosaurus brancai*, as a prominent part of the entrance to the institution. The array of dinosaur fossils in the diorama include a *Kentrosaurus* from the upper Jurassic Era, and these specimens are arranged to answer the question of what 'the world was like 150 million years ago' during the 'Saurierwelt' (world of the dinosaurs). The temporal distance between object and audience is thereby accentuated in this arrangement as the visitor is asked to witness the ancient era as a distant object from the perspective of the present. Whilst lacking in the same grandiose scale, the Burke Museum, the Washington State Museum of Natural History and Culture in the United States, displays the remains of the 'first dinosaur' specimen discovered in the state in the lobby of the institution:

> The 80-million-year-old fossil is a partial left femur bone of a theropod dinosaur, the group of two-legged, meat-eating dinosaurs that includes Velociraptor, Tyrannosaurus rex and modern birds. It was collected by Burke Museum paleontologists along the shores of Sucia Island State Park in the San Juan Islands in May 2012 (Burke Museum 2016).

The significance of the find, its age and its relationship to the ordering of time, space and place are paramount in this scheme. Whilst the find, a partial fragment of femur, was just over 40cm in length, its significance for scientific research is undeniable. However, its representation also provides a means by which the visitor can observe the passing of time and look upon the fossil from the perspective of an advanced and elevated position. In effect, this conforms to Bergson's (1910) refutation of scientific, categorised timeframes and his insistence on notions of duration (la durée). For Bergson, the systematic organisation of chronology did not provide for the 'sense of becoming' present both within individual humans and the wider environment. Bergson (1910: 29–30) critiqued Darwin's 'mechanised' model of evolution as a process which whilst correct in its wider scheme nevertheless obscured the role of consciousness: the élan vital (see Bergson 1908). It is these notions of la durée and the élan vital which were central to Bergson's perspective of memory and the manner in which past, present and future were conceived (Bergson 1910). It is la durée which offered a notion of time not fixed on the process of linearity, where order was reinforced and notions of advancement are essential, but where time is understood as an experiential phenomenon, a multiplicity where individual and social relations are formed. Bergson (1910) argued that time as an evolutionary concept had become spatialized, carved into discrete units which offered no concept of 'free will' or engagement (Deleuze 1988). As such, notions of 'endurance' are stressed here, as a means to convey continuance and association with the past and its lived sense in the present and future:

> Pure duration is the form which the succession of our conscious states assumes when our ego lets itself *live*, when it refrains from separating its present state from its former states. For this purpose it need not be entirely absorbed in the passing sensation of an idea; for then, on the contrary, it would no longer *endure*. Nor need it forget its former states: it is enough that, in recalling these states, it does not set them alongside its actual state as one point alongside another, but forms both the past and the present states into an organic whole, as happens when we recall the notes of a tune, melting, so to speak, into one another (Bergson 1910: 100).

It is this analogy of the 'tune' that enables Bergson (1908: 95) to define the significance of the élan vital, as a sense of existence within all living matter or inanimate
matter which had been shaped by a living force. It is this sense which connects
individual humans to one another and to the wider environment (after Minkowski
1970). The élan vital and la durée are linked by memory and the act of remem-
brance as lived experience, difference and alterity are forged (after Deleuze 1988).
In this manner, to observe the past is not to be overwhelmed or alienated by the vast
difference or chronological scale but to recognise and experience the points of
connection. In effect, it is not the sound of a fading or forgotten tune but the use
of those notes in a contemporary performance.

This consideration of Bergson's notion of time, matter and memory is not the
promotion of an elusive, ephemeral or esoteric concept of being, it is to assert
the significance of witnessing. To bear political witness to the temporal difference
between past and present serves to elevate the present and dispel the past; to
witness past and present as a social act, part of the same process, provides a point
of connection where the Latourian principle of the democracy of things can be
made apparent. This assertion of alternative time frames or visions of prehistoric
eras as part of the occurring and the imminent within modern society is certainly
part of exhibitions and displays across institutions. Within this space, visitors are
encouraged to regard time present rather than time past which opens up alternative
meanings as to how humans regard their own place in the environment and their
relationship to others who share that environment. This act of social witnessing
can be seen in the arrangement of dinosaur statues and models within museum
foyers or within the entrance grounds outside institutions. These replicas dominate
the interior and exterior of buildings and impress upon visitors the sheer size of
these ancient creatures. Some of the most notable examples of this type of engagement can be seen in the replica *Diplodocus* that was arranged outside the Carnegie
Museum of Natural History in Pittsburgh in 1999. This fibreglass model, fleshed
out and coloured in a brown-green hue, was erected to mark the centenary of the
discovery of the dinosaur in Wyoming. However, its placing outside the museum
attracted a degree of criticism and concern at the outset (Anon 1998). Indeed, the
local newspaper's editorial expressed concern that the appropriate level of behaviour
would be shown towards the model:

> The model took nine months and $200,000 to build. We hope that, along with
> fun and excitement, the dinosaur also generates a sense of respect, particularly
> given its proximity to two university campuses. College high jinks may be
> part of Oakland life, but this is one habitat that deserves an evolving reverence
> (Anon 1999b).

Indeed, such was the concern that over-excited visitors may attempt to hang from
the neck of the 25-metre model that the display was altered slightly, in opposition
to current palaeontological thinking regarding the posture of the animal, to raise
the head out of reach (Batz Jr. 1999). In this respect, the cast model, whilst affecting
an engagement between past and present, affirms the perspective of the political

witness where the object relays the chronological chasm that exists for the visitor. However, in the action of social engagement with the model, visitors have brought the dinosaur to life in some respects as it has moved from being a museum piece to a point of reflection and a site of congregation for both residents and visitors to Pittsburgh. Adorned with seasonal garb or the uniform of sporting teams, 'Dippy' has become an emblem for the city. With its own social media presence, the model would appear to represent a natural history which is lived with and its place within the urban and cultural landscape reflects its significance to the present rather than just to the ancient past. As such, the object is able to convey that sense of élan vital and la durée as it places individuals within the immediate context of a present and relevant past. This is not due to its status as reconstructed, 'life-like' model, but in its value as a point of engagement where time, place and memory are formed. This perspective has been utilised at the Naturmuseum Senckenberg, Frankfurt, Germany, where reconstructed models of *Diplodocus* and *Tyrannosaurus* are arranged for visitors to contemplate within a contemporary environment. This was also demonstrated in the 2003 initiative at the Carnegie Museum of Natural History named 'DinoMite Days' where fibreglass model dinosaurs were decorated by artists and installed around the Pittsburgh area (CMNH 2003). The collection of 100 models was eventually installed next to Dippy outside the institution before being auctioned by the museum to raise funds to renovate their dinosaur gallery. The event served to establish a point of connection for contemporary citizens:

> Pittsburgh will miss these wonderful creatures. Indeed, nothing like it has been seen around here since prehistoric times (Anon 2003).

Such examples of la durée within natural history are nevertheless somewhat limited as they are largely focused on present concerns. In effect, just as fossil remains may enforce a chronological separation, dinosaur models can induce a presentism that removes society from the engagement with and the social witnessing of the élan vital. Similar recreated dinosaur displays located outside natural history museums abound within the United States and across the wider world. However, these scenes could certainly not be regarded as innovative in practice as recreated, 'lifelike' sculptures have been used in public exhibitions since the nineteenth century. From the Crystal Palace dinosaurs of Benjamin Waterhouse Hawkins to the statue of the *Iguanodon* from Bernissart which stands on its hind legs outside the Musée des Sciences Naturelles de Belgique in Brussels, displays of models have been a consistent feature of the representation of natural history. The recent uses of reconstructions within the entrance halls and courtyards of institutions reflects this historic practice and also points to alternative uses.

At the Cleveland Museum of Natural History, a model of a *Stegosaurus* has been displayed at the entrance of the museum since 1968. The impetus for the display was derived from Sinclair Oil's 'Dinoland' display at the 1964 World's Fair in New York, where model dinosaurs were depicted in a 'natural' setting. A dinosaur model was commissioned from the same designers and arranged for display outside the museum. Provided with the nickname 'Steggie', the replica has become a treasured part of the local community by firmly establishing this ancient past in the present. Indeed, with an alteration in the model's colouring which was undertaken in 2016 in line with current knowledge regarding the extinct animal, the museum was able to reflect on its significance for their work (see DeMarco 2016).

As such, the display is always located in the present, designed to excite visitors to serve as social witnesses to the moment of their engagement with the statues. A similar process can be observed at the New Mexico Museum of Natural History and Science in Albuquerque, New Mexico, which was opened in 1986. Two bronze statues, named Spike (*Pentaceratops*) and Alberta (*Albertosaurus*), are on display outside the museum's entrance and appear to be staring at each other in advance of a clash between the herbivore and predator. Modelled after fossil finds from New Mexico, the display is intended to introduce visitors to the significance of the state's collection and emphasise the transformation of the environment of New Mexico since the Cretaceous Era. As such, it provides a means of reflecting on the past, present and future as the visitor is encouraged to engage with the display. Within contemporary museums these sculptures are used to engage and to educate, to establish wider connections between visitors and the ancient world. This parti- cular type of engagement is most evident within the Dinosaur Plaza constructed for the Fernbank Museum of Natural History in Atlanta, Georgia, in 2009. Designed as a public gathering space outside the institution, the tableau features *Lophorhothon atopus* arranged within a family group with details of the habits and behaviours of the creatures placed on a display board as well as information on their environment in Late Cretaceous Era Georgia. The bronze sculptures were thereby linked to the past, present and future for visitors as individuals, groups and communities were provided with a space to engage with the set of sculptures. What is emphasised in this arrangement is the connections between the immediate and the prehistoric:

> Now nearing completion, the life-size sculptures of a *Lophorhothon atopus* mama and her two tykes – she's 9 feet tall and 27 feet long, the juveniles (sic) half that size – will gather around the reflecting pool in the new entrance plaza. These dinosaurs roamed Georgia more than 60 million years ago, but our knowledge of them is relatively recent (Fox 2009).

> Causing double takes and a few hastily snapped photos by motorists, the bronzed creatures had been trucked . . . down I-75 and then Freedom Parkway, to their new reflecting pool home, Fernbank's Entry Plaza. But you could also say that the 24-foot-long mom, who commanded her own flatbed, and her two offspring have been winding their way here since the Cretaceous Period, when they lumbered across parts of Georgia. Paleontologists discovered their remains in the 1940s and recognized *Lophorhothon atopus* as a distinct species in 1960 (Langston 1960).

This 'domestication' of natural history, the use of familial connections as an organising principle within sets of dinosaur sculptures, is a common theme across museums. The rationale for explaining social behaviour through such displays is evident and it also provides a point of connection for contemporary visitors. However, it is in such connections that the alterity of the past could be lost as social and gender norms are reiterated within the arrangement of models depicting extinct creatures (after Machin 2008). Within these spaces, the individual acts as a political witness to the enactment of modern human social values through ancient settings (Haraway 1984; 1989). For example, within the entrance court of the Westfälische Museum für Naturkunde in Münster, North Rhine-Westphalia, Germany, an array of *Triceratops* sculptures was set out as a family group for visitors as part of the museum's redevelopment during the 1990s. Such displays provide points of congregation and engagement with the past but they inevitably transpose modern gender roles onto that past. Similarly, at the Royal Tyrrell Museum, Alberta, Canada, visitors are presented with a troop of *Pachyrhinosaurus lakustai*. This herbivorous species is well accounted for in the fossil record across Alberta and their location outside the main entrance of the institution provides for a point of connection between the ancient environment and the present day. However, the scene, which was part of the museum's redevelopment after its opening in the 1980s, portrayed maternal and paternal roles through the arrangement as adult males appeared to watch out over the group whilst adult females protected the young. Within such representations, natural history can cease to be something that is thought with as dominant modes of thought are asserted. Despite such strictures, notions of identity can be reformed within these spaces by the way in which natural history is offered as a part of the present. The prominence of dinosaur models located in the exterior spaces of museums provide points of reflection beyond the time frames installed within displays of fossils. Connected with a sense of the élan vital, the vision of alternative worlds and life can serve as a point of critique within contemporary society. As such, a democracy of things is initiated which can alter the shape of our own era.

It is not just dinosaur fossils and reconstructed models which emphasise time, space and identity within the courtyards and entrance halls of natural history museums. Within these locales are also the exhibition of prehistoric and modern animals, as fossils, lifelike models and skeletal displays, which function to provide particular witness perspectives for visitors. The prominent display of extinct fauna is significant in this regard as it affirms the chronological and evolutionary distance between the visitor and the object of their gaze (after Berger 1972). Mounted and exhibited as an emblem of an ancient era, such displays of Ice Age mammals or Pleistocene sea creatures emphasises the passage of time and the place of modern humans at the apex of development. It is within the reflection of the ancient, now non-existent and the evolutionary dead-ends that the projection of humanity's success can be affirmed. The view provided by such displays asserts a narrative of progression for which contemporary society is the conclusion and the political witness. This is essentially the model that Steinberg (1996) characterises as the 'how we got to where we are now'; a self-referential assessment which reinforces the significance and inevitability of the present (Berger 1972: 11). As such, removed from the contingency of history, these representations of the past are statements of the success of the present for visitors. They perform roles as evidence for an act of political witnessing where the ending of the narrative is already assured. Such engagements with the past limit the alternative ways in which the relationship between the past, present and future can be formed.

This narrative perspective can be observed within the display of prehistoric creatures within the plazas or foyers of natural history museums. For example, the Arizona Museum of Natural History in Mesa, Arizona, United States, which exhibits items and specimens which represent the natural and cultural history of the Southwest region, displays fossils of extinct animals within its Cenozoic Lobby. A mammoth (*Mammuthus columbi*), mastodon (*Mammut americanum*), pre-historic North American horse (*Nannippus*) and lion (*Pantheria atrox*) are displayed alongside other archaic specimens. The narrative effect of such arrangements emphasises progress and direction as visitors are able to witness the chronological separation from such species as well as their absence within the contemporary world. The habitats of these extinct animals within the local region or the connections of these ancient creatures to modern species only serves to reassert the distinction of humanity and the teleological inevitability of our own environment. A similar array of fossils and models from prehistoric animals can be observed in other national and local institutions. For example, in the entrance hall of the Museum of Natural History and Science in Cincinnati the skeleton of a mastodon (*Elephas americanum*) has been the central object of visitors' attention for nearly three decades. These pieces take on an iconic role within a museum, even becoming key features of institutional advertising, promotion and outreach work (see MacMahon 2013).

Such displays can be seen with the statues, sculptures or models of prehistoric animals displayed outside museum entrances. Indeed, the Museum of Natural History and Science in Cincinnati erected a family scene of four fibreglass mammoths in 1980 outside its initial site in the city. The display was intended to provide a definite vision of the past for contemporary audiences to observe (see Findsen 1980). This is also highlighted with the bronze mammoth displayed outside the University of Nebraska State Museum in Lincoln, Nebraska. Erected in 1998, this sculpture measures nearly 4 metres in length with trunk and front foot raised in an 'active' setting. Its plinth is used for information panels detailing the environment of Nebraska during the last Ice Age. The sculpture is modelled on the mammoth skeleton excavated in the 1920s which now stands on display within the museum's Elephant Hall. Given the nickname Archie, the museum's 'life-size' sculpture provides a fixed narrative of development from 30,000 years ago to the present day. In Paris, within the Jardin des Plantes, outside the entrance of the Galeries d'Anatomie Comparee et de Paleontologie, which is part of the Muséum National d'Histoire Naturelle, an anatomically correct sculpture of a mammoth alongside other prehistoric creatures is displayed to visitors. A similar scheme was used with the 'natural sculptures' of prehistoric animals, including a mammoth, outside the Montshire Museum of Science, Norwich, Vermont, United States, in 2015. This 'Prehistoric Menagerie', consisting of accurate represent-ations of a variety of animals from the Cenozoic Age, was initiated to introduce the idea of the success of mammalian species during this era. These displays evidence the use of a romantic, visual 'emplotment' for visitors where a process of self-identification in the narrative of the past to the present affirms humanity's role in the world (after White 1973). Political witnesses are formed which can testify to the place of the contemporary world as the inheritor of the past and the inevitable culmination of evolution and adaptation.

Skeletal models, statues and reconstructions of prehistoric animals are also present within the central points of institutions as structuring devices or exemplars of a museum's values and mission. These pieces, located in the main display halls upon entrance, serve to direct visitors towards an act of political witnessing that secures a fixed narrative of development. Whilst this undoubtedly possesses educational and scientific value in organising knowledge and framing the visitor experience, it can obscure the connections between past and present as well as the sense of the élan vital. In this rational translation of natural objects within displays, institutions have formed an objective translation of nature which is rendered into narrative space (after Duarte 2014). Within the Florida Museum of Natural History in Gainesville, Florida, United States, visitors are presented on entry with a mounted skeleton of a Columbian mammoth measuring over 4 metres tall. Unearthed within the local area and providing visitors with a point of connection to this ancient past, the specimen evidences the past environment of the region but also the cessation of this era and the eventuality of the present. The Nevada State Museum in Carson City, Nevada, United States, also displays the fossilised remains of a mammoth found locally in the Black Rock Desert. In this display, the animal is presented in the same manner that it is considered to have met its demise, falling into a water hole. In the Sam Noble Oklahoma Museum of Natural History in Norman, Oklahoma, United States, the Pleistocene Plaza adjacent to the Samedan Oil Corporation Great Hall is dominated by a bronze statue of an Imperial Mammoth (*Mammuthus imperator*). The sculpture, which has been present since the museum's opening in 2000, is the same casting as the mammoth sculpture at the University of Nebraska and is located with a backdrop of the museum's landscaped grounds. These displays, whilst engaging and interactive for visitors, nevertheless create particular narratives of time and place, affirming the act of political witnessing. The past is observed as a rationale for the present.

The display of extinct, prehistoric mammals is also used within museums to forge narratives of place and identity and to engage with a dialogue regarding the natural environment in the past, present and future. This can be observed in the display of specimens within entrance halls or foyers which create social witnesses of visitors to the institution. The placing of excavated mammoth bones from the local vicinity in exhibitions and displays is a particularly prominent aspect of this process. The Milwaukee Public Museum, Wisconsin, United States, is a clear example of this use of fossils to develop points of reference as its presentation of the Hebior Mammoth (*Mammuthus primigenius*) demonstrates a connecting point between the Ice Age landscape of the region and contemporary concerns. Situated in the atrium of the museum, the display of the mounted fossil cast is over 4 metres tall and the specimen was obtained from a 1994 excavation site in nearby Kenosha County, over 60km away. The institution has displayed this local find since 2007 and has used the evidence of butchery marks across the find as evidence of human settlement in Wisconsin from around 12,500 BCE. With the acquisition of the fossil remains, the museum's focus on the local and human context of the mammoth was emphasised in the local press (see Loohauis 2007).

Locating the mammoth in its place and utilising it as a means of emphasising local connections to Wisconsin's human settlement allows the object to be part of a wider narrative beyond its initial chronological context. This demonstrates the notion of la durée, the élan vital and memory as identified by Bergson (1908) as visitors are called upon to witness the fossil cast as something that is lived with rather than observed as a point of measuring a distant past. The excavated site in Kenosha County also yielded another mammoth skeleton which exhibited butchery marks and the cast of the fossils of this specimen are exhibited at the Kenosha Public Museum in Kenosha, Wisconsin. The skeleton of the Hebior Mammoth is erected as part of the institution's 'Wisconsin Story', an exhibition which narrates the development of the region's geography, climate, animals and people. The mammoth fossils are thereby juxtaposed from scenes of modern Wisconsin life which serves to connect the find with a sense of place. The evidence of human use of the mammoth acts as a plot device to consider past, present and future interactions between people and their environment (Kenosha Natural History Museum 2015). Such connections between specific places, people and the objects of natural history provide a mode of social witnessing which can serve to challenge and disrupt the present. The 2001 discovery of a fossilised skeleton of a mastodon in Simi Valley, California, United States, provided a point of reflection for local residents as the seemingly incongruous nature of an ancient animal and modern suburban sprawl was exposed. The fossils were studied by the Natural History Museum of Los Angeles County, where their display in 2010 as part of their 'Age of Mammals' exhibition reinforced this connection between the pre-historic world and modern life within the area through the mastodon skeleton and other specimens:

> The Simi Valley mastodon stomped around the area through which the Ronald Reagan Freeway now runs; the mysterious paleoparadoxiid was fossilized in ocean sediments forming modern-day Laguna Hills; and both the giant jaguar and saber-toothed cat stalked their prey right in the heart of what we now know as the Miracle Mile (Natural History Museum of Los Angeles County 2016b).

The Simi Valley local news also reflected upon this juxtaposition, which appeared to question the values of contemporary society and the impact of human society upon the area:

> With Simi Valley's modern-day landscape dominated by Mediterranean-style shopping centers, crisscrossing power lines and asphalt roadways, it's hard to imagine a prehistoric elephant lumbering around the area now intersected by the Ronald Reagan Freeway (Marsh 2010).

This point of association between places and natural history enables the connection of communities to their wider environment. Such responses can be noted in the relationship formed between the fossilised remains of an Ice Age mammoth and Ilford, Essex, Britain. The excavation of a Steppe mammoth (*Mammuthus trogontherii*) skull in the nineteenth century had caused great excitement as it represented the only complete cranium ever discovered in Britain. The fossil was taken to the Natural History Museum, where it has been exhibited and studied over the past century. However, in 2014 a campaign organised by charity groups and development agencies in the Ilford area began to call for the return of 'Monty' the Ilford mammoth. In 2015, this project was realised with the donation by the Natural History Museum of a resin cast of the object which was displayed as part of the 'Ice Age Ilford' exhibit at the local Redbridge Museum from 2015. This temporary exhibition, which also featured other fossils from Ilford that were part of the collections of the Natural History Museum and the Royal Geographical Society, connected visitors to the past environment of the local area, the Victorian scientists who discovered and examined the mammoth and the contemporary history of the wider borough. The role of natural history in these examples offers a point of reflection and dialogue between past and present. The fossilised mammoth bones are not regarded as an object of a distant age but an object of the contemporary world which has relevance and meaning for modern audiences. The assertion of chronological or evolutionary distance and development can divorce specimens of natural history from a living context. The eradication of the élan vital within displays and exhibitions prevents the formation of a democracy of things within the environment, the maintenance of a hierarchy of order and a sense of progress.

The arrangement of specimens, as a means to narrate the natural history of development and adaptation, serves to reiterate relationships with the environment in the present. This can be observed in the movement within some institutions to provide visitors with a demonstration of contemporary concerns. It is these concerns that were behind the decision to alter the display in the Natural History Museum in London from the cast skeleton of a *Diplodocus* to a Blue Whale. In essence, this marks a shift from a temporal or political witnessing to a moral witnessing where visitors are reminded of the pressing dangers faced by all species on Earth. Institutions in Europe and North America have employed this perspective to actively inform visitors of the contemporary natural world. This is usually demonstrated with the exhibition of models or reconstructions of extant species within foyers or main halls. Such displays could seem to offer a direct means of obtaining a 'democracy of things' as they appear to affirm the place of species alongside one another. However, it is not their concurrent existence with humanity that enables such an agenda to emerge. The effect could be formed through a display of prehistoric mammals or dinosaurs; what is significant is the use to which these exhibitions are mobilised. Certainly, the legacy of nineteenth century practices of collection and display have influenced this practice. The display of elephants, tigers and giraffes which had been hunted and mounted by colonialists served to exoticise, exploit and objectify the resources, peoples and landscapes of European possessions in Africa and Asia (after Said 1978). However, contemporary displays

Figure 3.3 The elephant at the National Museum of Natural History, Washington, D.C.
Carol M. Highsmith Archive, Library of Congress, LC-HS503- 2082.

are orientated towards conservation with some existing objects 'repurposed' for
this objective. Despite this alteration in outlook, these arrangements still require a
particular orientation in time, space and narration as they create moral witnesses to
the past and present.

Perhaps the clearest example of this is the African elephant (*Loxodonta africana*)
that still forms the centre of the National Museum of Natural History in Washington,
DC (Figure 3.3). The hide of the animal, which was donated by the explorer and big
game hunter Josef J. Fénykövi (1891–1971), was modelled onto a frame and
displayed in the central rotunda at the museum from 1959. Named after Fénykövi,
the elephant which once served as an object of wonder and excitement for its size
and presence has been redesigned over the course of the twentieth century to reflect
social, cultural and political alterations in the work of museums. It was most recently
reimagined in 2015 as an object of reflection and concern and a means by which
visitors can bear witness to the fascination of nature but also the need to protect such
species and their environment.

The new installation incorporates the latest scientific information about these
magnificent but endangered animals. Explore the evolution of elephants,

discover insights into elephant behaviour, and learn about the conservation challenges that threaten today's wild elephants. A new visitor information desk has also been incorporated to make visits to the museum easier and more enjoyable (SNMNH 2016a).

Similarly, the display of a specimen of an African elephant within the Schloss Rosenstein, the eighteenth century building which is today part of the Staatlichen Museums für Naturkunde Stuttgart, is used to detail the wider history of evolution on Earth. The Luonnontieteellinen Museo (Natural History Museum of Helsinki) also uses its taxidermied African elephant, located in its entrance foyer by the stairwell in its early twentieth century building, as a starting point for its wider exhibits on evolution and the diversity of nature. This is the objective of the Grande Galerie de l'Évolution in the Museum National d'Histoire Naturelle in Paris. In 1994, the nineteenth century exhibition space was redesigned with the central hall used to display the collection of animal specimens from across the world. These examples of taxidermy were formed from the collections that had been built up over centuries with the earliest animals being those which had been part of the menagerie of Louis XV (1710–1774). As part of the new exhibition, this collection was arranged into the animals obtained from Africa and organised into a grand procession across the main floor of the museum. Named as the 'caravane africaine', the display serves to demonstrate the ecological variety within the African continent and the significance of the natural world. On entrance to the gallery, the visitor is orientated towards this extensive tableau, which contains over 60 animals and is headed by an African elephant. The effect of such a display is to create an act of moral witnessing through the collection of taxidermy presented to the public. Rather than reflecting the colonial excess of European nations, the exhibition demands a recognition of these animals as a disappearing resource. This particular perspective was reasserted with the development of the galleries in 2014 which was covered within the metropolitan and national press (see Anon 2014).

As such, the Grande Galerie de l'Évolution becomes a site of moral witnessing, an action which is confirmed in the other wings of the museum. The 'Salle des Espèces Menacées et des Espèces Disparues' (Hall of Threatened and Extinct Species) within the Grande Galerie functions as a space of reflection where the fragility of the natural world is explored through the collection. Over 250 animals are presented to visitors within this darkened display room as individuals are called upon to reflect on the meaning of loss within the environment. This effect is engendered through the positioning of extinct species such as the Dodo (*Raphus cucullatus*) adjacent to those animals whose survival is under threat such as the Sumatran tiger (*Panthera tigris sumatrae*) (MNHN 2016). It is this use of nineteenth century taxidermy collections to provide exhibitions and displays which can also be noted in the work of the Manchester Museum, which is part of the University of Manchester, Britain. In 2013, the Victorian showcases of one of the main display rooms for taxidermy was redeveloped into an exhibition entitled 'Nature's Library'. Drawing upon the extensive collections of living, extinct and endangered species, this display challenges the action of cataloguing specimens by nineteenth century scholars by organising items on the basis of titles such as 'Symbols', 'Resources' or 'Experience'. These titles reflect the uses that humans have made of the natural world, emphasising the origins of products, medicines and ideologies in the natural world. It is this arrangement that fulfils the ideal of a 'democracy of things', as whilst the environmental history of human use and exploitation of flora and fauna across the millennia is explored, this is a narrative that asserts visitors as moral witnesses to natural history. This material and ideological connection between humanity and the wider environment places past, present and future as a point of consideration for those moving through the gallery. The division of objects into use and function does not rely upon the chronological order, evolutionary progression or taxonomic categorisation. Instead, what is referred to throughout is a sense of engagement and responsibility. In this manner, the élan vital of the environment is maintained and the taxidermied animals and products derived from animals are not rendered inactive. Through such representations, the moral witness is called upon to observe and to testify as to this account of environmental history, where the narrative is ordered not on progress but on an ever changing relationship.

The exhibition halls of natural history museums which exhibit the models, fossils, taxidermy and skeletons of extinct and extant species of the natural environment as part of a process of development and advancement assert the linearity of the past into the present. Whilst this undoubtedly accurately catalogues species and adaptation, it can deprive objects of their sense of existence. Reduced to mute tools which act as fixed points in a narrative of evolution, the engagement of visitors is as a means of acting as a witness to the triumph of enlightenment ideals. In effect, this is a form of political witnessing as exhibits affirm notions of place and identity within contemporary society. By orientating the visitor towards a grand narrative of progress, the points of engagement and reflection between the past and present can be obscured. This mode of organisation is prominent across museums in Europe and North America and serves as the *de facto* arrangement for institutions. It can be observed in the grand palaces to imperialism and industrialism which originated in the nineteenth century and it is utilised in the museums constructed in the last few decades to encourage local development and engagement. The ubiquity of this scheme demonstrates the success of this mode of organisation to dictate a comprehensible and comprehensive narrative but it also assumes a level of prominence that prevents alternative narratives. These modes of representation do not challenge the evolutionary scheme but demonstrate the way in which individuals, groups and communities can be separated from a sense of existence by frameworks that focus on categorisation and delineation.

The propensity of this arrangement can be observed across natural history museums despite the size of the premises or the scale of the collection. The Carnegie Museum of Natural History in Pittsburgh uses this scheme in its Cenozoic Hall which displays fossilised mammals as a means of detailing the history of evolution over the course of 66 million years. The display features a number of finds which were retrieved from the Agate Fossil Beds in Nebraska, United States, over the course of the first decade of the early twentieth century. These included the extinct mammals ranging from the *Dinohyus* (pig-like omnivores), *Menoceras* (ancient rhinos), *Moropus* (herbivorous hoofed mammal) and the *Menoceras* (ancient rhino). These animals, which were endemic to North America, are presented as part of a narrative of progress within the collection, moving from the archaic to the complexity of contemporary specimens. This is clearly identified with the display detailing the evolution of the horse which accounts for the development of the animal over the course of millions of years. The visitor witnesses the progress of evolution in such circumstances; the processes of adaptation are clearly defined as objects are arranged according to their place within a grand narrative of progress from early simplicity to contemporary success. Other archaic species from across the world complement this collection, with an Irish Elk (*Megaloceros giganteus*) and a Moa (*Megalapteryx didinus*). The extinction of these animals appears to affirm the inevitability of progress and adaptation. This is especially relevant with the museum's display of a mammoth, dire wolf, giant ground sloth and sabre-toothed cat, which were all retrieved from the La Brea Tar Pits, California, United States. This is a pattern of representation that exists elsewhere across other institutions.

As such, exhibition space can become a catalogue of motion, defining the inevitable movement of life through time. It is this notion of fixed time and space that was critiqued by Engels as motion divorced from matter (Engels 1894). In this manner, motion was never process alone but process which is born from, shaped and understood through material existence:

> Modern natural science has had to take over from philosophy the principle of the indestructibility of motion; it cannot any longer exist without this principle. But the motion of matter is not merely crude mechanical motion, mere change of place, it is heat and light, electric and magnetic stress, chemical combination and dissociation, life and, finally, consciousness (Engels 1940: 16).

Therefore, the display of motion is not only governed by the material existence of the animals themselves during their lifetime but by the material conditions of the visitors to the exhibition. Process and movement are understood through the perspective of the observer and, in the context of representations of adaption and progress through the millennia, the individual becomes a witness to the contemporary processes of capitalism as much as the development of early mammals on Earth. The assessment by Engels that Darwin's evolutionary theory affirmed a model of capitalist economics can be utilised to assess how representations of motion within contemporary institutions correspond to the Enlightenment ideals of progress and advancement. This is not to undermine evolution or the process of natural selection which was identified by Darwin; it is to acknowledge that the exhibition of such schemes can alienate individuals from the materiality of their existence and alternative forces of motion.

The division of time and motion in this manner can be observed in the use of halls and exhibition rooms that exhibit objects with the preface or the theme 'The Age of'. Whether dinosaurs, prehistoric mammals or Pleistocene sea creatures, groups are arranged by species, genus, family, order and class into evidence of a particular epoch. Perhaps the most commonly used mode of organisation for these displays is the 'Age of the Dinosaurs'. Under this heading, permanent and temporary exhibitions have been developed using a range of objects, from fossilised skeletons to animatronic devices. Whilst ever-increasing media are utilised to provide engaging and accessible displays to visitors, the organisation of the objects to classify a distinct period of time and era of development utilises the Darwinian-derived concepts of motion and advancement. In this arrangement, adaptation and progress are enshrined as universal where stasis, reversal and punctuated revolution could also be regarded as part of the evolutionary process (Gould and Eldredge 1977; Gould 2007). The past in constant forward motion as a mirror of the present serves to 'naturalise' contemporary ideas about development. As such, political witnesses are formed through such representations as individuals can attest to the form and function of 'western capitalism' in existence at all times and places. With the forces of the market shown as universal, the objects of natural heritage serve as emblems of the relationships, dynamics and development of contemporary society. Whilst providing a conceptual framework which illuminates complex concepts, it could serve as a colonisation of the past. The formation of a 'democracy of things' would rely on alternative notions of motion, progress and advancement.

This mode of organisation can be observed within the Yale Peabody Museum of Natural History in its Hall of Mammalian Evolution. The arrangement of specimens expertly details the processes of mammalian development during the Cenozoic Era following the disappearance of the dinosaurs after the Mass Extinction Event of 65 million years ago. The fossils of now extinct mammals such as the mastodon enable the explanation of mammalian adaptation to environmental conditions. Alterations in climate and corresponding changes in vegetation or predation can be observed as the catalyst in the diversity of mammals, from the archaic forms of elephants, rhinoceroses or horses to the modern-day equivalents. The sense of motion here characterised by Darwinian evolutionary schemes is clear. The drive for development through competition for resources alongside biological or physiological change can be observed as part of the ordinary state of nature. This is also present in the iconic and extensive mural created by the artist Rudolph F. Zallinger (1919–1995) which serves as the backdrop of the exhibition space. Entitled 'The Age of Mammals', the piece was completed in the late 1960s and reflected the close collaboration between the artist and the foremost scholars of the period. The mural depicts the rise of mammals in North America from the first warm-blooded creatures to the large Ice Age animals that dominated the continent up to 8,000 BCE. Reading the piece from left to right, the mural reflects the same processes of motion observable in the exhibition hall:

> These changes in the environment and vegetation have been skillfully woven into the tapestry of the mural itself, providing the viewer with a feeling for the forces that shaped the outcome of this vast evolutionary drama (Yale Peabody Museum of Natural History 2016b).

This is not to suggest that such conceptual schemes are erroneous but that the linear concept of progress represented within this type of display does constitute it as a place of political witnessing. The visitor to such sites can draw parallels regarding the 'natural order' of life on earth and their present economic circumstances. This does not have to act as a confirmation of existing frameworks and could act as a means of dissent against issues of economic inequality. Indeed, the organisation of past evolutionary schemes and its implicit correspondence to present socio-political and economic structures is one that Engels had originally noted could serve as a point of critique on the present:

> Darwin did not know what a bitter satire he wrote on mankind . . . when he showed that free competition, the struggle for existence, which the economists celebrate as the highest historical achievement, is the normal state of the animal kingdom (Engels 1894: 35).

The notion of development and improvement is regarded here as indicative of a society that valorises such ideals rather than an inevitable and essential aspect of the environment. It is within the display of human evolution that such organisational schemes have encountered the greatest criticism (see Scott 2006). The projection of contemporary agendas regarding race and gender upon exhibitions by both museums and visitors has been demonstrated by scholars (see Wiber 1998). This is a legacy of the racist and colonial frames through which human evolution has been understood from the nineteenth century onwards. Since the 1980s, prominent institutions have sought to address these issues with displays that explicitly challenge visions of 'primitive' and 'modern' humans (see Bennett 2004). This movement has resulted in displays such as the Anne and Bernard Spitzer Hall of Human Origins at the American Natural History Museum which opened in 2007. This original exhibition utilised fossils and genomic data to represent the discovery of humanity's origins. Featuring specific displays of early hominids such as *Homo ergaster*, *Homo erectus*, Neanderthals, and Cro-Magnons, the overarching theme linking the specimens is not about improvement and advancement. Rather, it concerns the mutual elements across the variations of hominids, namely notions of creativity which are presented as a uniquely human characteristic. This sense of a shared past and present is emphasised in connection with new and future discoveries which will further reveal the common human story:

> This hall is about all of us – about who we are and where we come from. Although the human family originated many millions of years ago, we know a great deal about our remarkable past. The rich human fossil record dates back more than six million years, and scientists are finding exciting new specimens all the time (American Museum of Natural History 2016a).

The Natural History Museum in London installed a new gallery for Human Evolution in 2015 along similar lines. This arrangement also drew upon fossil and genetic data to present a narrative of human evolution as an account of similar traits and shared heritage. Using hominin fossils, tools and reconstructions, the exhibition confronts visitors with the diversity of early hominid species and the way in which this past emphasises our common humanity:

> Embark on a seven-million-year journey, from the first hominins to the last surviving human species: us. Investigate what defines a hominin and how much we modern humans have in common with other human species, as well as what sets us apart. Along the way you will discover the changes in physical characteristics, diet, lifestyles and environments that have shaped modern humans (Natural History Museum, 2016a).

In 2004, the Yale Peabody Museum of Natural History (2016c) opened a permanent exhibition, 'Fossil Fragments: the riddle of human origins', which also focused on the shared issue by posing the question: 'where do we all come from?'. This query was answered by asserting the African origins of hominids and the way in which the development of *Homo sapiens* was not the inevitable consequence of the evolutionary process but the result of climate change and adaptation. The Naturhistorisches Museum Wien (2016b) in Austria uses environmental and cultural development as nonlinear modes of development in its recent display of the 'evolution des menschen'. In 2010, the David H. Koch Hall of Human Origins was opened at the National Museum of Natural History with a comparable mission of assessing 'what does it mean to be human'? In this gallery, on entry the visitor is guided through the timespan of human evolution, highlighting how the Earth and its environment has altered over the course of millions of years. Through reconstructed faces of the ancestors of modern humans and interactive family trees detailing the discoveries of hominid fossils across the world, visitors are encouraged to 'meet the ancestors'. The evolutionary model of progression and complexity is replaced with a focus on the human quality of creativity and alteration. With the exhibition's focus on environmental adaptation and human ingenuity, the displays attempt to chart the future as well as the past. As part of the closing display, the exhibition uses this narrative of humanity's evolution to address current issues of climate change and the use of natural resources. In the displays entitled 'Changing the World' and 'One Species Living Worldwide', the impact of humanity's spread across the globe is assessed in the context of its origins. As such, visitors are called upon to bear moral witness to a shared history which serves to highlight the challenges faced in the present (SNMNH 2016b).

This approach is also pursued by the concentration on biological processes as the driving element within hominid evolution within the Institut Royal des Sciences Naturelles de Belgique (2016) in Brussels. The Galerie de l'Homme, which opened in 2015, uses fossils and reconstructions to explain 7 million years of hominid evolution. The focal point of the gallery is the display of seven reconstructed hominids, including a model of a modern *homo sapien*. This 'impossible family portrait' is used to emphasise the complex and multidirectional nature of human evolution. Rather than asserting the rise from primitive to modern, the display asserts the diversity of humanity and the unifying nature of our physiology which was shaped across millions of years: 'notre évolution, notre corps' (Institut Royal des Sciences

Naturelles de Belgique 2016). The display of human evolution has certainly altered considerably within institutions across Europe and North America over the course of the last few decades as critiques regarding the misrepresentation of race and the promotion of gendered norms have been asserted (see Haraway 1989). Increasingly, museums are using the story of human evolution to address the legacy of racism and inequality that has marked previous assessments and representations of this history. The steady march of progress with hominid fossils moving from simple, hunched ape-like creatures to upright modern humans has been reassessed. Displays may focus on the development of bipedalism, such as the evolution of mankind exhibition at the Naturemuseum Senckenberg, Frankfurt, Germany. In this account, the upright gait of *Homo sapiens* is cast as the objective in the display of fossilised hominid specimens, but this is regarded as an adaptation to habitat and culture rather than representing the triumph of 'humanity' (Senckenberg Naturmuseum 2016). In this manner, the 'motion' of evolution has been reassessed and alternative models of development and 'progress' have been asserted.

This new mode of displaying human evolution is most clearly observed in the two European institutions which specialise in this field: the Musée de l'Homme in Paris and Museo de la Evolución Humana in Burgos, Spain. Both of these museums utilise a mode of representation that enables the visitor to witness homi-nid evolution as a shared history which draws humanity together. The Musée de l'Homme, part of the Muséum National d'Histoire Naturelle, begins its narrative of human development with a statement regarding diversity through a large array of busts capturing the likeness of individuals from across the world. This collection was formed during the nineteenth century and whilst originating within colonial ideals of ordering racial types is used within the museum's display to emphasise humanity's common attributes regardless of any physical difference (de Lumley 1998; Mohen 2004). In this manner, the Musée de l'Homme, which was founded in the 1930s and reopened after renovations in 2015, engages with the legacy of its collections. Focusing on physiology, biology, language and culture, the displays are arranged to engage visitors with the question of what it means to be human. Indeed, the museum is organised upon concerns of the origins of humanity, the identity of humanity and the future of humanity (Van Praët 2012). Through these structures of representation, it forms moral witnesses to our natural history as it requires visitors to acknowledge the heritage we share:

> We come from a very long evolutionary process as every other species, we are one species among millions of others whose appearance is, throughout the history of life on earth, very recent. But we are a peculiar species that thinks and thinks the world and, in fact, the change ... to a museum which it is subject and object (Musée de l'Homme 2016).

The Museo de la Evolución Humana, which was founded in 2010, serves as site of reflection through its display of finds of early hominids from nearby sites within the Sierra de Atapuerca. Within this mountain range the earliest hominids in Europe were discovered, including *Homo antecessor* and *Homo heidelbergensis*.

The museum presents these discoveries and guides the visitor through the processes of human evolution by focusing on physical and cultural development. From the importance of stone tool use to the significance of fire within early hominid development, the Museo de la Evolución Humana explains the process of change as one of cultural and biological adaptation. The institution uses this narrative to explain the qualities that humans share and as such casts itself as a site for understanding the past, present and future of humanity.

> It is not only a museum of our ancestors, but it is also a place to reflect on the present for our species Homo sapiens, endowed with abilities that allow us to change the world (Museo de la Evolución Humana 2016).

However, the organisation of exhibition halls and spaces focusing solely upon human evolution could serve to assert the perspective of the visitor as witnessing the unique aspects of humanity as distinct from the wider developments across the animal kingdom. This is reflected in the displays of some institutions which place human evolution alongside the evolution of the closest living relatives of *Homo sapiens*: chimpanzees, bonobos and gorillas. For example, the Staatliches Naturhistorisches Museum, Braunschweig, Germany, has located its display on human evolution adjacent to its exhibition on primate evolution to stress the common ancestry. The alterations in the manner in which hominid evolution is displayed has sought to emphasise common aspects of the human experience but could emphasise the unique status of humanity over other species. In a 'democracy of things', the presumption of the pre-eminence of *Homo sapiens* could emphasise a division between 'nature' and 'culture' (Latour 2004a).

In this manner, an ecological approach to evolutionary history could be considered which enables a recognition of humanity's place in the wider scheme of environmental development. As such, human and non-human could be regarded as interconnected parts of the wider narrative of evolution as well as the basis of a political ecology (see Latour 1998). Such schemes can be observed in museums where displays of evolution, which include human and non-human schemes, serve as a means of highlighting the way in which flora and fauna respond and adapt. For example, the Michigan State University Museum (2016) in East Lansing, Michigan, houses a 'Hall of Evolution' which details evolutionary processes from several hundred million years ago to the ending of the last Ice Age approximately 10,000 years ago. The array of specimens, from fossils of early lifeforms and a model of the fossilised remains of the hominin 'Lucy', which were discovered in Ethiopia in 1974 (*Australopithecus afarensis*), are arranged in a chronological, linear order. However, the positioning of objects alongside one another enables the visitor to witness a narrative that focuses on the human and the non-human alike. A similar theme can be found within the exhibition 'Minerales, Fósiles y Evolución Humana' which opened in 2016 at the Museo Nacional de Ciencias Naturales in Madrid. This display guides visitors through the development of dinosaurs, mammals and humans and culminates in the arrangement of minerals which are linked to their current use within society. As such, 'nature' and 'culture' are intermingled within the exhibition space as the three sectors demonstrate adaptation and change in relationship to one another (Museo Nacional de Ciencias Naturales 2016b).

This structure was at the centre of the permanent exhibition that opened in the Naturhistoriska riksmuseet, Stockholm, in 2008. This was entitled 'Den mänskliga resan' (The Human Journey) and detailed the development of human evolution over millions of years but also examined issues of human culture and physiology which are forwarded as means of adaptation. From the outset, visitors are encouraged to regard human evolution not as a process of achievement and success but as a many-splintered structure that saw variation and difference emerging in response to wider environmental conditions. Where the exhibition develops these concerns of and removes the focus on modern *Homo sapiens* as the culmination of the evolutionary narrative is within a concern for wider contexts. Whilst climatic change is examined, so too are the relationships built with other animals that enabled humans to exploit and survive. The exhibition is structured around specific points which are used to assert the place of humanity in the environment as part of a wider history of life:

1. Human evolution – a tangled bush
2. First on two legs
3. The first family
4. The great climate change
5. Teeth that can take it
6. Hand of the maker
7. The oldest tools
8. A body like ours
9. The true Europeans
10. Last but not least
11. From idea to reality
12. Our genetic journey
13. The human journey – the movie
14. The brain – a complex piece of gossip
15. Africa – cradle of human culture
16. Where do we go from here?
17. Mammal relationships
18. What is a species?
19. Elephants – not only giants
20. All the family
21. Walk like you
22. Island evolution
23. Horses in a changing environment
24. Monster birds
25. The rhythm of climate change
26. The changing climate in Europe
27. Woolly rhinoceros

28. Sabertoothed predators
29. Giant deer

(Naturhistoriska riksmuseet 2016)

The development of this mode of representing human evolution as a nonlinear process entwined with the changes that occurred within both the wider environment and other forms of life on Earth asserts the perspective of the moral witness. The visitor is called upon to bear testimony to the way in which humans have changed with others and the implications that has for current issues faced by humanity.

Displaying the resources of natural history

It is within the display of rocks, minerals and geological processes that visitors to natural history museums become moral, political and social witnesses. Indeed, the place of humanity in the world is formed through these representations. It is in the exhibition of the processes that have formed the environment and which structure our world that the connections between the human and the non-human are emphasised. It is within these displays that the ability of natural heritage to engage with notions of time and place are apparent. Whilst fossilised remains of extinct animals millions of years old may serve to emphasise the separation between past and present, rocks and minerals that are far older than these creatures are imbued with relevance for their current roles within modern society. In essence, this equates to the difference between observing specimens as indicative of a perception of time that can be termed endurance or perdurance (see Lewis 1986). That objects persist in the environment for millions of years is demonstrated by the rocks, minerals and metals which are exhibited in institutions of natural history across the world. Nevertheless, it is the nature of this persistence which is crucial in this definition as it alters the relationships that are built between humans and non-humans (Hawley 2002). The concept of endurance connotes the existence of an object throughout time as a single, fixed entity without altering its character. To bear witness to this definition of time is to observe these objects as continually part of the world. This is contrasted with the notion of perdurance, which argues that objects are not continual but composed of various states through time as they are observed and are part of wider society. The appearance of a fixed state of endurance might appear to be most relevant for describing the place of rocks, minerals and metals. The form and character of these objects appear to have persisted across millions of years which enables institutions of natural history to emphasise the deep chronological timespan separating visitor and object. However, increasingly museums have drawn upon a notion of perdurance which ascribes these objects with the ability to possess more than one singular character. In the display of perdurance, visitors witness the way in which the past is lived with and how it contributes to the present and future.

Museums of natural history have highlighted both the ancient origins and contemporary uses of rocks and minerals in displays over the last few decades as a means of developing accessible and engaging exhibits. One of the most prominent of these displays was the Harry Frank Guggenheim Hall of Minerals in the American Museum of Natural History (2016b) in New York. The exhibition initially opened in 1976 and with some additions remains a central part of the institution. Whilst the displays of rare, exotic and colossal specimens, such as a four-tonne block of azurite-malachite which was mined in Arizona, are prominent, it is the formation of minerals, metals and rocks and their use by human groups which present the past and present to the visitor. Exhibits from the California Gold Rush of the nineteenth century demonstrate the way in which individuals and communities have sought to mine, exploit and use the Earth's resources. As such, the display forms a powerful statement on the relationship between humanity and the environment. The Natural History Museum of Los Angeles County whose 'Gem and Mineral Hall' was opened in the 1970s and refurbished in the 1980s, also presents this connection to its visitors. Objects are presented here in their various guises, as building blocks of the Earth, evidence of colossal geological activity, as emblems of wealth and as motivators of human ideals and actions but as a consistent character or form (Natural History Museum of Los Angeles County 2016c). The 'mineralogie & petrographie' display hall at the Naturhistorisches Museum Wien (2016c) presents the unchanging nature of these objects as a consistent exploration of form, from building materials, minerals for chemical processes or gems for display. An act of political witnessing is formed with these points of interaction for visitors as the use of resources for human development is historicised and 'naturalised'.

The same mode of the representation of place and time can be observed within the Oxford University Museum of Natural History. Within its petrology displays in the rocks and minerals aisle, the museum provides guidance to the way in which these objects should be framed as 'British minerals: a heritage revealed'. Through the display of minerals and rocks from Britain and Northern Ireland, the museum engages the visitor in a dialogue regarding time, place and identity. In one section entitled 'Heritage preserved', the rare minerals which were once present in the caverns and mines of the country are displayed. In the comparative section, entitled 'Heritage revealed', the way in which common rocks and minerals are used for various functions such as construction within modern Britain is discussed (OUMNH 2016b). This representation of the endurance of these materials, emphasising the continuing properties of rocks, minerals and metals, serves to cast the visitor as a political witness. This mode of engagement is present within museums of various sizes from the national to the regional. For example, the Naturwissenschaftliche Museum in the German city of Aschaffenburg displays the minerals, metals and rocks of the Spessart region of central Germany. This exhibition uses the materials to demonstrate the important of mining and extraction for industry in the area (Naturwissenschaftliche Museum 2016). The observance of human use and exploitation of materials as an ongoing process ensures individuals can testify to the normalisation of this action in the present and the future. However, a point of departure with this process can be noted in the displays at the Hillman Hall of Minerals and Gems at the Carnegie Museum of Natural History, Pittsburgh. This hall, which was originally opened in 1980, exhibits over 1000 pieces to illustrate the formation, character, value and use of minerals within society. The latter was of particular concern in the exhibition's redevelopment in 2007 where new sections detailing how particular minerals had shaped histories, cultures and identities was included. From the use of tin within Britain, minerals from Romania or silicates from India, the exploitation of resources is linked to the development of regions and nations in the past and the present.

Where notions of perdurance are made present within displays of geological and mineralogical specimens, the relationship between humanity and the wider environment can be reconsidered as inherent values and properties are reimagined and alternative ideals are negotiated in the contexts of wider, global environmental challenges. This can be most clearly seen in the 'Hall of Minerals, Earth and Space' at the Yale Peabody Museum of Natural History which opened in 2008. Within this permanent exhibition, meteorites are used to explain the origins of the solar system, mineral specimens indicate the tectonic forces that shaped the world and geological displays evidence the atmospheric and climatic alterations that have taken place since the Earth's formation. Significantly, the rocks, ores and minerals that constitute the exhibition are presented as part of an ongoing issue, not as objects fixed in purpose and function for human exploitation. Indeed, the perspective of a moral witness is formed within the exhibition as the visitor is required to assess the impact of human action on the wider environment. With displays on climate change, visitors testify to the way in which humans and non-humans interact. The objects are presented in the manner of perdurance; rather than possessing some stable form in all times and places they are shown to be part of a wider ecological system which is represented by constantly changing physical and ideological states:

> The earth we live on is a dynamic, ever-changing planet. Geologic forces operating over billions of years have shaped the earth we know, and our planet will continue to change for billions of years to come. Today, societies depend on the earth's vast mineral and energy resources, as well as on a habitable climate that allows life to flourish (Yale Peabody Museum 2016d).

As such, time, place and identity are not innate and universal categories for rocks and minerals, they are culturally and chronologically specific attributes that are acquired through engagement within wider social networks. These issues are explored in the Museum für Naturkunde in Berlin, which houses one of the largest mineral collections in Germany within its traditional nineteenth century display cabinets. Whilst the items on display are classed by place and arranged taxonomically within their original museum contexts, new cabinets are used to assert a different account of these materials to demonstrate the multifarious identity of these objects:

> Precious metals such as gold, silver and platinum have been traditionally valued for their use in jewellery and minting. In addition, they have now become essential in medicine and catalyst technology. Glass and ceramics are indispensable in our high-tech world, and computers could not run without the element silicon (Museum für Naturkunde 2016a).

Therefore, the notion of *perdurance*, the ability to possess multiple places, times and roles, is observable in such schemes. It is within the display of minerals, rocks and geology and that this particular form of engagement appears so prominently. Over the last two decades, as institutions have developed their displays on 'Earth Sciences' in response to environmental concerns and the awareness of humanity's impact on the climate, an experiential, phenomenological mode of representation has been assessed (see Schneider 2016). Certainly, the way in which visitors to natural history museums are encouraged to consider the forces that have altered the planet in the past, present and future does place the individual sense of being as the measure of things. This can be seen in the 'Volcanoes and Earthquakes' permanent gallery which opened in the Natural History Museum in London in 2014 (Natural History Museum 2016c). The exhibition guides visitors through the themes of volcanoes, plate tectonics and earthquakes, providing geological specimens to support the data and describing how human cultures have lived with these natural phenomena over thousands of years. The focus on experience is evident in the earthquake simulator which is the central feature of the exhibition. This interactive piece replicates the physical experience of being caught in an earthquake. The scenario provided for visitors is the setting of a supermarket in the Japanese city of Kobe, which suffered an earthquake in 1995. However, whilst the display may vividly engage visitors with the effects of nature, it also provides a distinct sense of social witnessing where humanity is not isolated from the history or future of the planet but intimately connected.

Earthquake simulators have become part of the process of engagement for contemporary natural history museums. They serve as a powerful means to inform visitors of the effects of seismic activity but they also enable a point of connection with this natural history which informs the present. In 2009, the Natural History Museum of Crete (M.Φ.I.K.) put in place its own 'educational seismic table' to provide a physical demonstration of the type of natural events that frequently occur within the region. The natural history of earthquakes is detailed and the human engagement with these events in the past and present is provided (Natural History Museum of Crete 2016). As such, whilst the human emotions and physical sensations are part of the display, what is affirmed is the act of social witnessing. The simulator itself is designed to represent a school room which emphasises the point of connection between the human and the non-human. This process of socialisation, which challenges the binaries of natural and cultural, is part of a wider European programme to address the emotional trauma of earthquakes for children (RACCE 2010). The same scheme supported the later installation of an earthquake stimulator at the Natural History Museum of the Lesvos Petrified Forest in Sigri on the Greek island (Μουσείο Φυσικής Ιστορίας Απολιθωμένου Δάσους Λέσβου). The same setting of a classroom was used in the displays and schools and social groups are encouraged to utilise the recreation as a means of preparing for earthquakes (Petrified Forest of Lesvos 2016). The museum places the simulator firmly within the wider geological history of the region, with evidence from the rocks and minerals from Lesvos used to detail how these events are part of the area's history. The visitor can also use the monitors that provide readouts of current seismic activity in the eastern Mediterranean and across the wider world. Within these exhibitions, the visitor becomes a social witness to how earthquakes are part of the past, present and future of the planet and their destructive force is regarded as a threat that is lived with.

In the Californian Academy of Sciences in San Francisco, the permanent exhibition 'Earthquake: Life on a Dynamic Planet' also teaches preparedness in a region prone to significant seismic activity. Visitors are presented with the notion of earthquakes as part of geological processes that shaped the planet. As part of the interactive displays, individuals can experience the 'Shake Room'. This mimics the effects of an earthquake of a magnitude of 6 to 7 on the Richter Scale in the surroundings of a Victorian parlour. The use of such interactive pieces to relay the ecological aspects of geological events was explored earlier by the Field Museum in Chicago. In the temporary exhibition, which still tours the United States and Canada after its initial display in 2008, entitled 'Nature Unleashed: Inside Natural Disasters', the focus was on ensuring that the earthquakes and tidal waves which commit such destruction are part of a wider global system which includes humanity. The exhibition enabled individuals to create a seismic effect and construct volcanoes on simulation programmes but the engagement with a sense of social witnessing was paramount. Natural phenomena such as earthquakes and typhoons were presented as part of human culture and society. The regular occurrence of seismic activity across millions of years was displayed in the exhibition alongside the history of human tragedy which accompanies such events as well as the history of human engagement with earthquakes, tornadoes and volcanoes. These events are highlighted as experiential but it is not the human experience of these occurrences which is forwarded as significant. Newspaper reports of the initial exhibition highlighted this value of the display (Mullen 2008).

It is the ecological experience that asserts humanity's place in a world composed of non-human elements. Natural disasters are thereby regarded within the context of perdurance as a means to socialise not humanise nature and assert how it is the way in which such events are lived with that marks our society. To be a social witness to this natural history is to comprehend it as a human and non-human point of concern and association. Potentially, this radically alters relationships with the environment as the natural heritage is regarded as a set of altering concerns, not as a predefined set of qualities (Latour 2004a). Whilst it may appear that displays of geology have become too focused on engaging the senses, it is beyond the individual and on the social world where these exhibitions demonstrate their function. The representation of time, not as a fixed point but as a fluid negotiation of place and identity, where the same object or event may possess multiple meanings, enables this 'political ecology'.

The way in which astronomy and planetary science are represented in museums provides for a further demonstration of the role of natural heritage in establishing a sense of time and place. Within the portrayal of the formation of the universe, the solar system, the sun and the planets, institutions provide a means of moral witnessing. Through exhibitions, displays and the use of planetariums, the natural history of the universe's origins and development provide visitors to an institution with a means of regarding their place and their role within a society composed of humans and non-humans. Indeed, it is the reflection of the scale of humanity in relation to the history of the formation of galaxies, nebulae and stars that might appear to render *Homo sapiens* redundant. However, in the scope of such vast time-scales, transformations and distances, individuals are called upon to testify their connections to the world, not to the insignificance of their own being in the face of such sublime sights. In effect, the division between notions of endurance and perdurance can be noted here within institutions, as whilst the former presents the solidity and fixity of the universe, the latter presents the apparent stability of eons as a product of a fluctuating set of relationships (after Hawley 2002). The representation of endurance within the display of planetary science can be regarded as part of an established tradition which provided a means of fixing time and space to enable observation. This mode of engagement provides for an act of moral witnessing in its presentation of the universe for the gaze of the individual; a sense of ownership and control is contained within this engagement which places humanity as a measurer not as an element within this process. Through the arrangement of time and space through perdurance, the logical coherence of solidity is not abandoned, rather visitors are provided with a means to witness a sense of place within a wider sense of past, present and future. To be a witness to the formation of galaxies, stars and the planets to observe the processes of the universe, its development and the shaping of our own world within the solar system is to establish an association with a unique part of the natural heritage.

The portrayal of the perdurance of the universe is a key feature of the modern natural history institution. Situating visitors as participants in the constantly moving and altering states emphasises the socialization of this natural heritage. Humans and non-humans are presented in this array as part of the same system. The representation of planetary science as an ongoing event can be observed within institutions in Europe and North America. The Arthur Ross Hall of Meteorites at the American Natural History Museum in New York demonstrates this process of witnessing time and place. The display, which opened in 2010, focuses on the development of the solar system, tracking over 4.6 billion years of history. To explain this process, meteorites are used as demonstrations of the constantly changing nature of the solar system. As evidence of the forces that formed the Sun and its orbiting planets billions of years ago and as elements of planetary change within the present, meteorites are represented as possessing multiple states at multiple times. In effect, the display allows visitors to witness these artefacts as elements of the past, present and future. The display is divided into three areas which explain the origins of the solar system, the formation of planets and the reformation of those planets through meteorites. The exhibition links the occurrence of meteorite activity across the solar system with the evidence of meteorites on Earth including retrieved fragments and models of impact sites. Interweaving accounts of human engagement with non-human elements, the processes of an ecological politics can be observed as meteorites are accorded with agency:

For thousands of years, people have been fascinated by streaks of light flashing across the night sky. These 'shooting stars' are actually tiny grains of dust from space that burn up in Earth's atmosphere before reaching the ground . . . . they contain vital clues that help scientists understand how stars like our Sun formed and how planets, including Earth, took shape more than four billion years ago (American Museum of Natural History 2016a).

This is particularly the case with the colossal fragment of an iron meteorite weighing over 30 tons named Ahnighito which was removed from Greenland in the 1890s for display at the museum. The piece, which is the centre point of the exhibition, is estimated to be almost as old as the Sun and is used to illustrate the cosmic processes that have shaped our world but also how humans have interacted with this aspect of the environment. Time and place are thereby demonstrated as qualities that are made in conjunction with others, not measured as concerns which are in and of themselves. Visitors witness the formation of our solar system and can regard it not as a distant primordial event but as a process which reflects and impacts upon their engagement with the world.

In 2013, the National Natural History Museum opened its temporary exhibition 'The Evolving Universe', which served as a clear example of this use of perdurance. Within this display of photographs and panels, the notion of being multiple places in multiple times was detailed to visitors as a means of engaging with notions of vast scales of change:

> The farther we peer into space with powerful telescopes, the farther back into the history of the universe we see. The light from our Sun – a mere 93 million miles away – takes only a few minutes to reach Earth. But when we look at stars and galaxies in the night sky, we are seeing light that has travelled for millions – even billions – of years to reach us (SNMNH 2016b).

The physics of light, space and time enable the notion of perdurance and serve as a point to witness the cosmic events as occurring within an environment known to the visitor. The same point is explored within the permanent exhibition entitled 'Kosmos & Sonnensystem' in the Museum für Naturkunde in Berlin. In this exhibition, which was installed in 2007 on the stairway in the institution, meteorites are displayed surrounding an audio-visual system where the visitor is immersed into a presentation which establishes a relationship between the origins of the universe and the present:

> Let us take you on a fascinating journey through time and space – from the Big Bang to the present. A mobile sky projection disc floats in the 12-metre-high stairwell of the Museum, introducing visitors to the history of the universe. The multimedia show explains the formation of the universe and the Earth and briefly introduces all other planets of the solar system – Mercury, Venus, Mars, Jupiter, Saturn, Uranus, Neptune, and, of course, the sun itself (Museum für Naturkunde 2016b).

It is in such displays of planetary science that a conception of a democratic nature can emerge, where humanity is not the measure but a component in a wider system that is dependent upon the relationships that are built within it. The exhibitions of the universe and the solar systems form a means of building a sense of social witnessing; where a sense of society is constructed from a range of participants beyond the human. It is this mode of representation which is also present in the planetariums that are part of natural history museums. These displays provide a further means for visitors to establish a sense of place beyond the confines of the museum itself and form associations and a society with the processes that formed the wider solar system. The shows and exhibitions which are held within these arenas serve as points of connection regarding place and identity. For example, the World Museum in Liverpool, Britain, uses its planetarium to deliver programmes which negotiate space, time and life. Within shows held in this display room, such as 'From Earth to the Universe', the negotiation of a vast scale and chronology is undertaken through an assessment of humanity's interaction with the changing cosmic array (World Museum 2016). Similarly, the Cleveland Museum of Natural History (2016a) uses its planetarium (the Nathan and Fannye Shafran Planetarium), which was installed in 2002, for shows that connect the individual to the wider processes of the universe, emphasising the connections and similarities between Earth and elsewhere.

This sense of immersion is a key feature of the natural history within the planetariums of institutions. Individuals are placed within this greater celestial process and required to consider their place in it. In effect, the natural history of the universe is mobilised to regard the relationships that exist within society. The representation of this heritage under the notion of endurance, whilst fixing the object to be observed, can be regarded as revision of the flux in which these objects exist. The representation of perdurance within these sites provides for a framework which adjusts visitors to a wider sense of society and being. This can be observed in the way in which these parts of natural history museums are presented to the visitor. For example, in the Museum of Natural History at the University of Michigan, the creation of a sense of place is made paramount:

> These immersive films place you at the center of the action, as stars explode around you, and galaxies swirl about. The sense of motion and depth can be overwhelming, exciting, and beautiful (Museum of Natural History, University of Michigan 2016).

The Santa Barbara Museum of Natural History, which houses the Gladwin Planetarium, opened originally in 1957, forwards the same objective of experience and engagement but this is not offered as purely sensory. The apparent tendency towards a phenomenological display is not an exercise in solipsism but an engagement with a 'socialized' universe:

> You can tour the constellations of the night sky, fly through the Milky Way galaxy, explore the surface of the Moon, and ride a rockin' roller coaster on Saturn's moon Titan! (Santa Barbara Museum of Natural History, 2016)

Equally, the Naturhistorisches Museum Wien (2016a) provides visitors to their planetarium with an assessment of time, space and place. The planetarium provides a space of moral and social witnessing as individuals are required to understand their lives in the context of how other lifeforms have developed within Earth and potentially elsewhere in the universe. The Naturhistorisches Museum utilises this perspective in their planetarium show entitled 'Origins of Life':

> Origins of Life deals with some of the most profound questions of life science: the origins of life and the human search for life beyond Earth. Starting with the Big Bang, in chronological order, the show deals with the prebiotic chemistry in the Universe, the formation of stars, formation of solar systems, and the first life on Earth . . . as well as our search for . . . life beyond planet Earth (Naturhistorisches Museum Wien 2016a).

Across institutions in Europe and North America, the representation of astronomy and planetary science is asserted as a universal part of the natural history which offers visitors a means to 'transcend' their lives on Earth. This is not purely for the purpose of entertainment or for individual pleasure, it is frequently offered as a means of rethinking humanity's status and its relationship with the wider world. The social witnessing of such processes within the planetarium provides a means for redefining identity, place and politics within society.

Conclusion

The museums and institutions which represent natural heritage play a key part in the formation of politics, place and identity within the modern world. Indeed, it is within these sites that individuals can build associations with others, both human and non-human, that define humanity's role within a broad concept of society. Through the display of sculptures, models, fossils and reconstructions, the point of connection between humans and natural history is made. The formation of natural history in motion, in stasis or in connection with other things can serve as a means whereby other visions of the past, present and future are considered. These museums and institutions are places of witnessing, where moral, social or political visions of society are formed. As such, museums structure the concepts of connection between the ancient and the modern within the scope of natural heritage. The recognition of both the significance and the malleability of natural history is vital for building new ideas in society to respond to present day concerns. Natural heritage has perhaps been removed from the scope of a critical heritage studies for its appearance as timeless and inevitable. These epistemological claims belie the assumptions that order the representation of natural history (after Haraway 1989). The demonstration of time, social order, economics, gender and politics within displays and exhibitions removes the appearance of scientific neutrality as the connections to the wider world are exposed. In essence, what is required here is not the abandonment of the scientific but its inclusion within a far wider scheme of the social (Latour 2004a).

The aims and missions of institutions that seek to inform and educate the populace demonstrate the value of natural heritage as a significant part of national, cultural or social identity. However, the establishment of authority and expertise can obscure the ways in which visitors might engage with this heritage but also the manner in which that heritage is represented. Whilst arrangements of sculptures and fossils of dinosaurs or extinct mammals within the entrance halls of institutions may orientate the individual towards the past as a distant object of contemplation, it can also alter the manner in which the environment is perceived in the present by that individual. To regard natural heritage is to place oneself in relationship to the processes that have shaped the entirety of the planet, the solar system and the universe itself. This affords the opportunity of understanding humanity as not the observer, the measurer, the guardian of the natural world, but a participant among other non-human participants. The recognition of this society does not abandon the scientific process, it recognises that it takes place in a context with others. This is reflected in the representation of minerals, geological phenomena and planetary science within institutions where a recognition of how such objects are part of a wider social world can enable alternative discussions regarding time, place and identity. This is the ecological politics that Latour (2004a) identified as necessary with regard to the current environmental challenges that are present on Earth. It is within the study of natural heritage that origins can be understood but it is also within this heritage that the present and future of humanity can be reformed.

4 Sites of natural history

Introduction

It is through the creation of a 'sense of place' that the natural history within protected landscapes and conservation parks has been presented in western society. These sites have been reserved as significant locales where the heritage of flora, fauna, geology and hydrology have been regarded as important to the nation state. Whether mountain ranges, valleys, gorges, moorlands or ancient forest, these places are regarded as preserving a sense of nature which is threatened by human development. Whilst museums and education centres may provide guidance and information within protected parks, the places themselves are represented as offering an engagement with natural heritage. As such, they form a key part of the formation of a political ecology (Latour 2005). However, the status of national parks in Europe and North America as 'protected' creates the potential engagement with these sites as locked in a temporal and social stasis. Such uses of these spaces conforms to Nora's (1984) distinction of milieu de mémoire and lieu de mémoire. The former represents an invention, an attempt to recreate a tradition to preserve order and value. Milieux de mémoire are designed to address the loss caused by the forces of modernity which have removed the lieux de mémoire, the pristine places where society was connected to the environment, place and time. In this manner, a sense of place is created but it is one which is entirely fabricated, dependent on forming an imagined vision of the past. However, such nostalgic reveries prevent an understanding of how a sense of place is formed beyond the binary notions of 'tradition and modernity' and 'nature and culture'. Within the parks, centres and reserves that have been created within the contemporary western world over the past century, a mode of social witnessing can be made present which can radically alter notions of what constitutes 'society'.

In-situ locales of natural history provide points of witnessing how humanity is part of a wider network of others, both human and non-human. As such, these spaces create a distinct sense of place for individuals, groups and communities who use them. Social and moral witnesses are formed through the representation of geological features or natural processes which connect past, present and future but also demonstrate the role of a broad-based understanding of society. This is not to say that an aspect of this socialization is not present already within the national

Figure 4.1 Glacier National Park, Montana. Carol M. Highsmith Archive, Library of Congress, LC-DIG-highsm- 04925.

parks and conservation areas in Europe and North America. Indeed, this was a principle of their founding, to serve as a means for nation states to recognise key components of their identity within the natural environment. These concepts are still present within the operation of these areas. The focus of the preservation of places as a means of establishing national character or identity is an aspect of this socialization of nature. The National Parks Service (NPS) in the United States, celebrating its centenary year in 2016, draws upon this notion of heritage to represent national parks as the physical and ideological bedrock of the United States (Figure 4.1). For example, in the 2015 assessment of this natural resource, the NPS asserted the vision of leadership and management of sites as an undertaking for the nation:

> America's identity is rooted in our diverse geologic features, incredible landforms, and expansive landscapes. Natural features ranging from a single cave stalactite to an iconic landform like Delicate Arch to entire mountain ranges such as the Sierra Nevada are part of this identity. These striking features, compelling landforms, and quintessential landscapes are all part of what is called America's geologic heritage (NPS 2015).

However, the focus on society here is concerned purely with the human; wider concepts of the non-human are reduced to symbols evoking an imagined past as part of national narrative (Anderson 1983). In effect, this perpetuates the natural/cultural divide within the environmental movement. The representation of natural heritage within these sites can serve as more than a passive backdrop where humanity asserts itself and its values. National parks, conservation areas and protected wildlife zones serve as places where such distinctions can be negotiated as a 'democracy of things' is formed. Rather than appealing to some ideal of a pristine wilderness, or regarding places with values of character or nationhood, these sites demonstrate how natural heritage forms a key part of humanity's past, present and future as a participant within society.

Across Europe, Canada and the United States, nature conservation areas and national parks frequently cast their role as a form of protection against the incursion of modern life (see Gissibl, Höhler and Kupper 2012). Natural heritage is preserved in these places as if it is a static element where change would alter its character and purpose (Sellars 1997). This is enshrined in both legislation and the mission aims of the various services responsible for operating these sites (after Meringolo 2012). However, other narratives and modes of representation are also present within locales of natural heritage and provide both points of affirmation and a means of contestation (after Fairclough 1993: 41; 2001: 26). Indeed, the manner in which these sites are formed as a sense of place can be characterised on the basis of three types of engagement:

- A discourse of protection – formed through the formal pieces of national legislation and international agreements which define how a site of natural heritage is treated and perceived within wider society.
- A discourse of origins – these are formed from the representation of the sites themselves by government bodies, charities and private groups. They provide a frame through which natural heritage is regarded.
- A discourse of experience – these are formed from the engagement with places and landscapes on the ground through walks, trails and guides which connect individuals, groups and communities to the wider environment.

Through these elements, individuals, groups and communities bear witness to the place and purpose of natural heritage. The 'discourse of protection' can be identified through the legislation, mission aims, and documents of the national and international bodies responsible for administering parks and reserves as they frame and present natural history and heritage to wider society (after Mels 2002). Through this mode of representation, a distinct act of political witnessing is observed as national parks are presented as locales where natural history must be protected, preserved and shared as a means of cultivating common ideals within society (after Goggin 2012). Similarly, the discourse of origins, which emerges through the portrayal of these sites and landscapes by local authorities and non-governmental bodies affirms a sense of moral witnessing where a sense of reverential care is required for natural sites in order that these ancient formations be preserved from wider human society. However, the discourse of experience offers the potential for an alternative vision of the relationship between humanity and the environment.

Legislation and natural heritage: a discourse of protection and place

The discourse of protection can be identified in legislation which has enabled the development of national parks. This acts as a subtle but powerful form of engagement as society is structured on these frames of reference. In Britain, the fifteen national parks were all formed through the 1949 National Parks and Access to the Countryside Act. This legislation was amended with the Environment Act of 1995 which provided for the 'conservation and enhancement' of areas of natural beauty, wildlife and cultural heritage in England and Wales (Environment Act, 1995: Section 61(1)(a)). This mission was also accompanied by a second provision in Section 61(1)(b) which called for the promotion of opportunities for the wider public's development of understanding and ability to enjoy the 'special qualities' of those areas (Evans 1992: 227). Significantly, within this act, the notion of a natural and cultural divide is maintained. The areas of natural heritage which are conserved are contrasted directly with public access. The distinction is clear in that natural heritage is presented as an object that is continuously imperilled by human groups whose access to it is limited to momentary incursions into the areas that possess 'special qualities'. This provision in the legislation is the formalisation of what was known as the Sandford Principle, named after Baron Sandford (1920–2009). Sandford's role from 1971 to 1974 as chair of the National Parks Policy Review Committee saw the formation of a policy that clearly divided human and non-human engagements with these spaces:

> National Park Authorities can do much to reconcile public enjoyment with the preservation of natural beauty by good planning and management and the main emphasis must continue to be on this approach wherever possible. But even so, there will be situations where the two purposes are irreconcilable . . . . priority must be given to the conservation of natural beauty (Sandford 1974).

This particular type of discourse frames sites of natural heritage as places of political witnessing where individuals can regard the division between humans and the wider environment as inevitable. The genre established by legislation frames natural heritage as an element that society should consider apart rather than part of a wider society. This 'discourse of protection', where natural heritage is viewed as an element away from human society, is present within the legislation protecting national parks within Europe and North America (see Bastmeijer 2016). Indeed, it serves as a common aspect of western society's engagement with their natural heritage. In the course of attempting to secure the preservation of natural heritage in the form of geological formations, ancient landscapes and indigenous flora or fauna, the form of this mode of representation appears to remove notions of human engagement or bar the inclusion of this environment within wider concepts of society.

The creation of Yellowstone Park across the states of Wyoming, Montana and Idaho in 1872 by an Act of Congress demonstrated the divide between culture and nature that has persisted within the conservation, preservation and environmental movement. This site was selected for protection to prevent the incursion of humans who were regarded as ensuring the exploitation and ruination of sites of nature. The legislation which ensured its protection focused on two aspects which have defined humanity's engagement with natural heritage in the modern era. The first is the separation of people from the environment and the second is the notion that such action ensures the survival of this place in perpetuity. The 1872 Act states that the area be:

> ... set aside as a public park ... for the benefit of the people ... and all persons who shall locate or settle ... shall be considered trespassers (United States Congress 1872).

Whilst attempting to secure the integrity of the area, this legislation establishes a distinct sense of place for the territory. It is an area set aside from society which whilst providing benefit for the body politic by virtue of its existence is always apart from that society. This is the 'discourse of protection' which strictly divides culture and nature to the extent locales of natural heritage are regarded as places beyond the material aspirations of humanity. This was certainly part of the development of the Banff Hot Springs Reserve which was formed as Canada's first national park in 1886. Under the legislation which brought this site under public ownership, the potential disruption caused by competing claims to exploit the springs as a commercial venture provoked the Canadian Government into action (PC 2197, 25 November 1885). In both contexts, the threat of unrestricted human access to the places of nature was regarded as damaging and deleterious to any future society. As such, the 'discourse of protection' establishes the perception of these places as different and affirms the distinction between culture and nature. At the outset, these sites are formed by and themselves form an act of political witnessing. Within these protected areas visitors do not observe the untouched wilderness, capturing the scene of an earlier age, they observe the manner in which the state guides its populace and attempts to establish the direction of the future. It is within these places that society divides nature from culture and assumes the mantle of modernity (after Latour 1993).

The establishment of the National Parks Service (NPS) in the United States in 1916 can also be regarded as a formative point in the development of the 'discourse of protection' and the definition of areas of natural heritage as a distinct sense of place that should be preserved. The formation of the NPS was undertaken to ensure that the national parks that had been formed after Yellowstone in the late nineteenth century were appropriately managed to prevent exploitation for commercial interests. Within the 'Act to establish a National Park Service' (known as the 'Organic Act') (National Parks Service Act 1916), the key features of the modern drive for conservation are established. The act defines areas which are set aside from human exploitation for the express purpose of ensuring that they are maintained for the nation's progress and progeny:

> The service thus established shall promote and regulate the use of the Federal areas known as national parks, monuments, and reservations hereinafter

specified by such means and measures as conform to the fundamental purpose of the said parks, monuments, and reservations, which purpose is to conserve the scenery and the natural and historic objects and the wild life therein and to provide for the enjoyment of the same in such manner and by such means as will leave them unimpaired for the enjoyment of future generations (National Parks Service Act 1916).

It is this mode of witnessing places of natural heritage which has dominated the modern era. This constitutes a distinct act of political witnessing as it requires the separation of nature and culture whilst calling for the colonisation of the future with the values of the present (Giddens 1990). Under the protection of the National Park Service, areas within the United States were protected from development, which enabled the preservation of flora and fauna, though the desire to maintain the historic natural environment was defined by the status of these places as emblems of modernity. The same contrasting processes were present within the Wilderness Act of 1964 which remains a central tenet of managing natural heritage within the United States. This legislation, which strengthened the ability of the state to protect the environment of sites, also provided a means by which areas of natural significance could be defined in the future. The 1964 Act ascribed significance to the areas which were 'untrammelled by man' and where 'man himself is a visitor who does not remain' (The Wilderness Act 1964). The separation of humanity and natural heritage is made explicit in this legislation as a sense of place within the locales of natural history is defined as an area apart from society. The Wilderness Act possesses and promotes the key features of the discourse of protection that characterises the modern era: separation from humanity and preservation for contemporary and eventual society. The threat posed by modernisation, in the form of population explosion, urbanisation, industrialisation and environmental pollution, is a retreat to a nostalgic vision of the past for specific sites. Places of natural history are simultaneously formed by those two competing tensions, caught between history and the future; what is required from these places is endurance and stability in the face of change. The Wilderness Act attempted to address the growing concern that economic development had brought the despoliation of the nation's natural heritage:

> In order to assure that an increasing population, accompanied by expanding settlement and growing mechanization, does not occupy and modify all areas within the United States and its possessions, leaving no lands designated for preservation and protection in their natural condition, it is hereby declared to be the policy of the Congress to secure for the American people of present and future generations the benefits of an enduring resource of wilderness (The Wilderness Act 1964).

This forms the basis for the legislation in the twentieth century that defines sites of natural history – it is an attempt to seal off and preserve an imagined memory of a place. Within the discourse of protection is a clear separation of nature and culture which ensures that specific locales are regarded as apart from the normal processes of humanity. This is not to forward the notion of exploitation and development on sites of special environmental interest which preserves an important legacy, it is to acknowledge that an alternative relationship with natural heritage can be formed through legislation. Indeed, it is through legal processes that a new sense of place can be imagined for these locales. As such, concepts of a wider society which includes sites of natural heritage as more than the object of conservation and concern are prevented by the legislation that defines these places. Natural heritage is made central and significant in the modern state but it is regarded as separate.

The same mode of witnessing is utilised within the legislation which organises national parks in France (see Howard and Luginbühl 2015). The Act of 22 July 1960 which established the premise for national parks across the regions was based on the division of areas of the country as beyond the engagement of human society (see Larrère 2009). This distinction is enshrined in the legislation which requires the witnessing of these areas as separate:

> The territory of all or part of one or several communes may be classified by decree in the Council of State as a 'national park' when the conservation of fauna, flora, soil, subsoil, air, water and, in general, a natural environment of special interest and importance for preserving the environment against all natural effects of degradation and remove any artificial intervention likely to affect the area's appearance, composition and evolution. The area defined by the decree may also extend to maritime domains (Loi n°60–708 du 22 juillet 1960).

The notion of 'artificial intervention' essentially refers to human engagement with the environment. As such, we witness not the democracy of things but the delineation of particular zones of interest. The 1960 Act provided for the creation of the first designated area of protection, Vanoise National Park in Savoie. This was followed by further parks across France and in French dependencies including Réunion and Guadeloupe. In total, 10 parks have been created with the most recent being Parc national des Calanques in southern France. Each of these sites has been defined by the legislation set in place in 1960 which firmly distinguishes the human from the non-human. Within the terms of Article 3 of this legislation there is provision for the classification of boundaries of core and periphery within the national park. The core area is deemed as reserved for enabling the development of flora and fauna as well as protecting geological formations. However, the periphery zone is defined as accessible for the benefit of economic, social and cultural development which has enabled a range of tourism initiatives, exhibition centres and educational resources to have become part of the exterior landscape of national parks. Whilst enabling engagement with the spaces of natural heritage, the division of areas of influence, where humans can become part of society and where they cannot, does emphasise the appearance of a rift between 'nature' and 'culture'.

The operation of the national parks in France has been revised over the decades with consecutive acts or regulatory codes that have succeeded in reinforcing this 'discourse of protection'. Within this legislation, the areas have begun to be defined specifically as 'le patrimoine culturel', essentially the nation's cultural heritage. Whereas the 1960 Act served to demarcate areas for protection, later legislation accords areas of natural history as significant for the state to preserve in its entirety for its relevance to a sense of national identity (see Loi n° 2006–436 du 14 avril 2006). The effect of this legal framework is to create political witnesses to a particular definition of the significance, value and the relationship between humanity and the environment. 'Le patrimoine culturel' establishes the natural heritage of the parks as an example of pristine wilderness that must be left untroubled by human presence. This notion is also summarised in the Environmental Codes of 2016 which define the effect and the level of protection in France for national parks (Article L331–1 2016). This discourse frames the areas of national parks as an object removed from society for the betterment of that natural heritage. Whilst the need for protection against exploitation and pollution is clear, the establishment of boundary, both physically and ideologically, between humans and non-humans whilst regarding 'nature' as representative of an untroubled concept limits the potential of a democracy of things. The relationships that are built within a wider society are obscured in this manner as legal frameworks set forward national parks as a site of political witnessing.

The operation of legislation within Europe continues this division between culture and nature which casts natural heritage as an unassailable object rather than as a participant within society (see Wurzel 2002: 1–2). The Italian legislative framework that governs the operation of national parks demonstrates this process. The 24 Italian National Parks were largely designated as protected areas during the 1990s with the earliest site of Parco Nazionale del Gran Paradiso in Piedmont being assigned special status in the 1920s. The protection of national parks is provided for in the Italian legal system under the framework defined by Legge 6–12–1991 n. 394 which details the protection offered to the natural environment to prevent their development or pollution by the actions of humanity:

> . . . to guarantee and promote, in a coordinated manner, the conservation and enhancement of natural heritage of the country (Legge 6–12–1991 n. 394).

Significantly, within this law, natural heritage is accorded significance for its value to the state: del patrimonio naturale del paese. Whether a national park, regional park or nature reserve, the legal rights accorded to the organisations protecting these sites under n. 394 observes the danger and threat which the natural environment is under. Legislation protecting sites of natural heritage existed within the Italian legal system prior to the passing of n. 394. Indeed, the new post-war Italian constitution of 1948 established the nation's role in safeguarding 'the natural landscape and the historical and artistic heritage of the Nation'. However, n. 394 served to clarify the status of protected areas but also to establish the relationship between the human and the non-human within the state. Natural heritage is regarded as an object to be admired and preserved in its entirety rather than as an active element within society.

Within Spain, the legal framework for protecting areas of natural heritage first emerged at the outset of the twentieth century (Eritja et al. 2011). The first national parks were formed during this era in northern Spain – the Ordesa y Monte Perdido National Park and the Parque Nacional de la Montaña de Covadonga (known now as Parque Nacional de Picos de Europa). After the end of General Franco's military dictatorship, the new democratic Spain included a provision regarding conservation within its 1978 Constitution. Under Article 45.2, the state was invested with the guardianship of the environment as a resource for the people:

> The public authorities shall watch over a rational use of all natural resources with a view to protecting and improving the quality of life and preserving and restoring the environment, by relying on an indispensable collective solidarity (Constitución Española, 1978).

In this manner, natural heritage is deemed necessary to protect as part of ensuring the health and well-being of the state. Whilst replicating the ideological distinction between nature and culture, such a position does regard how the social, physical and cultural well-being of human society is linked to the wider environmental processes. Despite such social connections, it is the role of the political witness which is evoked within this framework as Article 45.2 requires the observation of natural heritage as separate. Although the Constitution put forward this recognition of the environment and its significance for the people, the legal protection offered to areas of natural history has only been clearly defined under legislation passed in the last two decades. For example, this principle is present within Law 4/1989 of 27 March, Conservation of Natural Areas and Wild Flora and Fauna, which enabled the development and preservation of 'natural parks' within Spain (Ley 4/1989). These areas of conservation were set aside from the economic development of the state to provide for the fulfilment of the principles set down in Article 45.2. One of the other key pieces of this framework of protection was the passing of Law 42/2007, of 13 December 2008, on Natural Heritage and Biodiversity. This defined the ways in which the environment and the natural heritage of the state was under threat from human activity. The law clarified a series of key points which concerned the way in which areas were to be protected but it phrased these objectives within the context of the relationship between the state and the environment as a source of enrichment for citizens, both socially and politically:

> In today's society there is a significantly increased concern for the problems of the conservation of our natural heritage and our biodiversity. The globalization of environmental problems and the growing awareness of the effects of climate change; the progressive depletion of some natural resources; the disappearance, sometimes irreversible, of many species of flora and fauna, and the degradation of natural areas of interest, have become a serious concern for citizens. They are claiming their right to a quality of environment to ensure their health and well-being (Ley 42/2007).

The definition of natural heritage as 'patrimonio natural' provides a means for society to regard this environment as more than just aesthetics or conservation for conservation's sake. Rather, it characterises this heritage as a means by which the quality of life for citizens is enhanced and preserved for future generations. In essence, the preservation of the natural world is forwarded as a key component of the modern state. This is most noticeable in the national constitutions that emerged in the aftermath of the break-up of the Soviet Union after 1989. Within these declarations of independence, democracy and modernity, 'natural heritage' and the preservation of sites of natural history were placed as central to the nation (see Żylicz 1994). This was in part due to the ecological damage sustained during the period after the Second World War and the protection of natural heritage became a key part of the independence of Eastern European states (Żylicz 1994). Indeed, the constitutions of many these nations have specific conditions set aside for preserving natural heritage (see May and Daly 2014). The 1992 constitutions of Estonia and Lithuania both draw upon the environment as a founding principle of the relationship between the state and citizens. The Constitution of Lithuania places this responsibility on the state so its enforcement can be witnessed by the populace:

> The State shall take care of the protection of the natural environment, wildlife and plants, individual objects of nature, and areas of particular value, and shall supervise the sustainable use of natural resources, as well as their restoration and increase (Constitution of Lithuania 1992).

In the context of Poland, one of the earliest responses to achieving autonomy and addressing the legacy of environmental damage was the implementation of the 1991 Nature Conservation Act. This legislation provided for the establishment of sites of protection, which has seen the proliferation of national parks, nature parks and conservation areas across the country. Significantly, the 1991 Act, whilst classifying the types of areas that were to be protected, also posits the ideal relationship between citizens and the environment that should be worked towards. The notion of a 'społeczny opiekun' (social guardian) was forwarded under Article 49 of the 1991 Act whereby every citizen of the Polish Republic was invested with the potential power to protect nature and educate the wider populace. The departure from a previous economic and industrial exploitation of natural heritage established a new relationship with the environment. This can also be observed in the Polish Republic's constitution of 1997 which also demarcated sites of natural history and the wider environment as part of 'dziedzictwa narodowego' (national heritage). This founding document of the Republic used the association between citizens and natural heritage and framed this connection to cast the latter as an endowment to enrich the body politic. Indeed, Article 5 of the constitution connects this concern to the very fabric of Polish national identity:

> The Republic of Poland shall safeguard the independence and integrity of its territory and ensure the freedoms and rights of persons and citizens, the

security of the citizens, safeguard the national heritage and shall ensure the protection of the natural environment pursuant to the principles of sustainable development (Konstytucja 1997).

The employment of 'social guardians' as a means of engagement between society and natural heritage as well as the constitutional and legal recognition of this as an ideological, economic and cultural asset to the nation has altered the way in which these post-communist societies are ordered in regard to the environment. In effect, this forms individuals, groups and communities as political witnesses to the significance of natural history. The very concept of the national identity is tied to the environmental heritage of a place. To witness the places that are preserved and protected is to regard the structures of the state. This is a significant feature of the 'discourse of protection' which can be observed across these examples; it is the way in which protection and conservation are interlinked with the ideals of the modern nation state (after Giddens 1990). The 'colonisation of the future' is activated through this legislation as it not only preserves places but preserves particular ideas of the relationship between the human and the non-human. Natural heritage is kept in a pristine state, which serves to enhance the environment, but the connections between this context and human society is stalled and limited to established traditions rather than open to change and development. This serves to 'naturalise' the relationship between human society and natural heritage as one where the latter appears neutral, devoid of meaning or passive and removed from participation which is constantly under the influence of the former. The legal frameworks define the manner in which natural heritage is regarded and how it forms part of society.

However, a different concept of human and non-human interaction is provided within a number of alternative legal contexts. Sweden has a long tradition of legislative protection for sites of natural history. The first national parks were formed by law in the early twentieth century which enabled the considerable protection of wilderness areas, essentially maintaining the natural heritage of the nation by minimising the ability of human society to impact upon areas deemed to be of national significance for their apparent unspoilt appearance (Prop. 1909). For example, the site of Hamra nationalpark in Gävleborg, central Sweden, was one of the first sites created to prevent the incursion of humanity to exploit the natural resources of wildlife and timber in the area. Whilst a system of protection was in place, this was altered in 1964 with the introduction of Naturvårdslagen (Environmental Protection Act). This legislation was significant as it described how national parks, nature reserves and national monuments were to be protected from development and exploitation. Whilst the division between nature and culture is present within this act, there is a recognition of a shared vision of natural heritage:

Nature constitutes a national asset to be protected and nurtured. . . . Everyone must show consideration and caution in their relations with nature (Naturvårdslag, 1964).

Natural heritage is thereby cast as a mirror through which society measures itself. Individuals are formed as moral and political witnesses as they are required to consider how their interactions with the environment alters the world. This perspective is similar to the later description found within the Polish legislation of citizens becoming 'social guardians' of natural heritage. This does mark a subtle distinction within Sweden, as the Naturvårdslagen provides the basis for a new relationship within the Swedish legal system which reflects a sense of co-existence with natural heritage. This is reflected in Section 13 of the 1964 Act, where 'natural monuments', areas of environmental distinction such as rock formations, water courses or landscapes, were included both for their unique status and aesthetic value but also because of their use by human communities (Naturvårdslag, 1964). It is this particular relationship and a recognition of the associations between the human and the non-human which underlies the basis of more recent legislation protecting natural heritage in Sweden, the Heritage Conservation Act (Kulturmiljölag 1988). Within these provisions, ancient human settlements and natural formations were regarded similarly as objects worthy of protection and classed together as 'ancient monuments'. This legislation, which has been modified repeatedly since its enactment to address emergent issues, also reasserts the relationship between citizens and the environment as being a key part of society:

> It is a national responsibility to protect and nurture the cultural environment. Responsibility for cultural heritage shared by all. Both individuals and authorities to show respect and care towards the cultural environment. Anyone who plans or carries out the work must ensure that damage to the cultural environment is avoided or limited (Kulturmiljölag 1988).

The legislation within Sweden which protects the natural heritage of the state was brought under one single piece of legislation in 1998 with the passing of a rationalised system of conservation outlined in the act known as Miljöbalken (Environmental Code). This set forward a key principle of 'modern' human and non-human interactions with Sweden's natural heritage: the notion of 'sustainable development'. This shift in perspective enables an alteration in how natural heritage is viewed, not as a site which is preserved in its entirety, but as a place where a wider participation with the environment is acknowledged. Miljöbalken did not impact upon the areas of natural heritage such as the national parks which remained removed from human society but it did acknowledge a dynamic relation between the human and the non-human and a focus on the occupation of a shared space:

> The provisions of this Code are to promote sustainable development so that present and future generations a healthy environment. Such a development is based on the insight that nature is worthy of protection but that the right to modify and exploit nature is a responsibility to manage nature well (Miljöbalken 1998).

Through this legislation, citizens become political witnesses to an altered state between individuals, communities, societies and the state's natural heritage. The vision of a society of things is absent in this arrangement, but it does acknowledge that the natural heritage is not beyond the scope of human engagement and exists within a common system with humanity. This alteration in how natural heritage was regarded and the movement towards concepts of sustainable growth was the product of wider international movements from the 1970s onwards. One of the first international conventions organised by multinational initiatives that examined how the natural heritage of the planet was managed was in the United Nations Conference on the Human Environment which was held in Stockholm in June 1972. Within this meeting, representatives from nation states examined the way in which the global shared environment could be protected. The focus on global concerns rather than national interests reflected the growing sense within environmental legislation that natural heritage be regarded as a common element across the world. The declarations that emerged from the conference highlighted a need for rethinking the relationships between human and non-human to address the growing threat posed by environmental damage and industrial pollution. The 1972 conference sought to address the problems faced by late twentieth century society by establishing a principle of moral witnessing towards natural heritage. Whereas previous national legislation had focused on preserving nature and removing it from human society, the declarations within this United Nations conference focused on the relationships that were formed with the heritage of the natural world:

> Man is both creature and moulder of his environment, which gives him physical sustenance and affords him the opportunity for intellectual, moral, social and spiritual growth. In the long and tortuous evolution of the human race on this planet a stage has been reached when, through the rapid acceleration of science and technology, man has acquired the power to transform his environment in countless ways and on an unprecedented scale. Both aspects of man's environment, the natural and the man-made, are essential to his well-being and to the enjoyment of basic human rights: even the right to life itself (UNEP 1972).

In this manner, the 1972 conference placed humanity as political witnesses to the destruction of the environment and invested within society with the demand that the natural heritage of the world be preserved. This recognition was explicitly made for both present-day society as well as for future populations. The relationship that is formed with regard to natural heritage therefore becomes a central feature for the modern world. Indeed, it is forwarded within the 1972 conference as a means to achieve the continuation of humanity on Earth. The subtle but distinct shift in how humans and non-humans interact as part of a wider society is evident within the conference's findings, as it describes how humanity has a 'special responsibility' to safeguard and manage the heritage of the natural world (UNEP 1972). Places and objects of natural heritage are thereby regarded not as places removed but as potential sites of human engagement. This perspective was given further definition a few months after the Stockholm conference with the General Conference of the United Nations Educational, Scientific and Cultural Organization (UNESCO) meeting in Paris during October to November 1972. This meeting recognised the global significance of natural heritage and the importance of its preservation to humanity. Indeed, it was through this conference that the status of 'World Heritage' site as a place of global significance was defined and the modes of preservation were agreed amongst international signatories. Declaring both natural and cultural heritage to be of the same value and deserving of the same level of care, the UNESCO (1972) conference put forward a sense of moral duty or righteous cause that the 'protection, conservation and presentation' of natural heritage for future generations was carried out. As such, there is a shift in how humanity's engagement with the natural environment and its legacy is discussed as the declaration of the conference calls for participation and engagement with natural heritage as a means of protection rather than calling for the wholesale preservation of natural sites and places. Article 5 of the declaration called for an integration of the environment within society rather than its removal:

> To adopt a general policy which aims to give the cultural and natural heritage a function in the life of the community and to integrate the protection of that heritage into comprehensive planning programmes . . . (UNESCO 1972).

The reports of both conferences in 1972 set forward an alternative approach within the preservation and management of natural heritage. National governments responded to the calls made within these declarations and began focusing their legislative efforts on working towards concepts of sustainability and development. International cooperation on these issues was also accelerated which further defined the relationship between humanity and natural heritage. Within Europe, cross-national collaboration on issues of protecting the environment brought forward binding treaties within the member states of the European Community to protect natural heritage. In 1979, the Convention on the Conservation of European Wildlife and Natural Habitats was signed in Bern, Switzerland, by members of the Council of Europe. It has since been ratified by the European Union, its member states and some nations within Africa. It forms the basis of heritage protection within Europe and illustrates the contemporary regard for how this heritage and the sense of place it induces is central to understanding the modern world. Whilst the Bern Convention, as it has become known, provided frameworks to increase protection levels for plant and animal life across national borders, it also framed a particular relationship between humanity and natural heritage that relied upon an act of political witnessing. This convention requires citizens to observe the significance of natural heritage in order to preserve the present and future of individuals, groups and societies across the world:

> Recognising that wild flora and fauna constitute a natural heritage of aesthetic, scientific, cultural, recreational, economic and intrinsic value that needs to be

preserved and handed on to future generations (Convention on the Conservation of European Wildlife and Natural Habitats 1979).

This notion of the environment as constituting part of a 'common heritage' which must be preserved for the benefit of 'future generations' has become an integral part of European legal frameworks on conservation (see Hunter and Smith 2005). Indeed, it is the basis of one of the earliest pieces of legislation regarding the environment issued by the then European Economic Community, the 1979 Birds Directive. Within this directive, migratory birds become a means of establishing transnational ideas regarding natural heritage as the common elements within European society. This act of political witnessing transforms the vision of natural heritage into a shared sense of place where values and ideals can be regarded as part of a programme of solidarity. The 1979 directive states:

> Whereas conservation is aimed at the long-term protection and management of natural resources as an integral part of the heritage of the peoples of Europe (Council Directive 79/409/EEC 1979).

The basis of the European programme of protection for natural heritage appears to have been attempt to promote wider integration across the continent and as such was influenced by existing systems of protection (see Weale 2005: 128). One of the most active European states with regard to environmental protection from the 1970s has been Germany (Voghera 2011: 107). A range of protected sites today exist within Germany, including 16 national parks and nearly 100 nature parks (see Lekan 2004). The national parks have emerged over the course of the twentieth century through various initiatives but they are all defined by the Gesetz über Naturschutz und Landschaftspflege (Bundesnaturschutzgesetz [BNatSchG]) (Law on Nature Conservation and Landscape Management). This legislation was initially enacted by the former West Germany in 1976 and revised over the decades with its last major amendment occurring in 2009. This law set out the importance of conservation and protection of the environment but it also considered the way in which human society and the various elements that constitute natural heritage interact and engage. These interactions were not specifically defined to designated sites and spaces. Indeed, the provisions laid out within this legislation recognised the significance of natural heritage within populated and non-populated areas. What lay at the basis of this legislation was a recognition that these places of protection were significant because of their interaction with humanity, not because of humanity's absence from such spaces. This recognition was born out of the wide-ranging definition given to what constituted 'natural heritage':

1. The performance of the ecosystem,
2. the usability of natural goods,
3. the flora and fauna and
4. the diversity, uniqueness and beauty of nature and landscape.

(BNatSchG 1976: 3574)

The original format of the legislation in 1976 also forwarded the concept of mutual cooperation between the environment and humanity regarding the stability of this relationship as essential in securing both the protection of sites and the economic benefit of society. Significantly, this acknowledged the potential for a changing relationship between humanity and the wider environment. In essence, the presumed stability and continuance of 'nature' is disputed as what constitutes natural heritage is shown to be the product of social engagement:

> The objectives of nature conservation and landscape management shall be implemented according to the principle of precaution and sustainability. This requires a proactive and conceptual approach to the work ahead. All alterations to nature should also be evaluated for their long-term effects (BNatSchG 1976: 3574).

As such, this defined sites of natural heritage as places that are part of national society and public life rather than areas which should be sealed off from human engagement. Indeed, the framework provided by the BNatSchG ensures that cooperation is the only means by which security and stability can be achieved for both the human and the non-human within Germany. The revisions to the legislation in 2009 reaffirmed this principle with the assertion that natural heritage should be maintained for the purposes of 'die künftigen Generationen' (the future generations) (BNatSchG 2009). It is on this agenda that the German context for the preservation of places of natural heritage was so influential for the development of European structures for conservation (Wurzel 2004).

The European legislation invests natural heritage with the ability to transform the perception of citizens of the continent. Whether landscapes, natural sites or wildlife, the legal protection of these spaces served to create a specific locale which reflected wider political objectives. In effect, natural heritage is cast here as part of the modernist project of the European Economic Community as it seeks to define and create the future (see Giddens 1991). Sites of protection, places of conservation, are not attempts to preserve the past in a pristine condition, they are the setting of how societies might be organised in the course of time. Natural heritage is the basis of how individuals, groups and communities might be reformed. Whether as a transnational project and a means to shape the future, the legislation defining places of natural heritage did alter during the 1970s and 1980s as a focus on a sense of sharing access to resources was formed in response to the dangers of environmental degradation. This process of alteration within the legislative frameworks can be observed with the World Commission on Environment and Development, which was founded in 1983 by the United Nations Secretary General and which published its findings in a comprehensive report known as the Brundtland Commission in 1987. This document defined the natural environment as the 'common heritage of mankind' and brought forward the concept of 'sustainable development' as a recognition that previous associations with natural heritage which had sought to manage the use of resources had not brought stability to a global population:

The time has come to break out of past patterns. Attempts to maintain social and ecological stability through old approaches to development and environmental protection will increase instability. Security must be sought through change (World Commission on Environment and Development, 1987).

The established division regarding the human and the non-human is still present within the report, but notions of a 'common heritage' and the recognition that the relationships which define ideas about the environment and society are malleable provided a new impetus for legislation across Europe from the 1990s onwards. This was also encouraged with the landmark agreement achieved in 1992 at the UN Conference on Environment and Development (UNCED) in Rio de Janeiro, Brazil (The Earth Summit). This conference resulted in a mutual agreement to address environmental issues and to develop programmes for 'sustainable development'. This corresponded to an alternative vision of natural heritage which behoved an integration of the interests of the human and the non-human:

Human beings are at the centre of concerns for sustainable development. They are entitled to a healthy and productive life in harmony with nature (UNCED 1992).

The definition of natural heritage within the UN Conference in Rio as a shared resource whose careful management could support sustainable economic, social and political development marked an international agenda in altering society through an engagement with this history (Panjabi 1997). In essence, this emphasises the manner in which a new sense of a global society was designed to be fostered through an engagement with natural heritage. In this new era, humanity would regard the environment as a shared legacy but one in which an altered sense of engagement would be formed which sought balance rather than exploitation. Sites of natural history, therefore, become key components of how the modern world is formed and reformed. The Earth Summit in 1992 set forward a vision of the future through the framework provided by natural heritage. The divide between nature and culture was still present in this image of how societies in the next millennium should be structured, but it is the redefinition of the relationship as one of connection and engagement that reflected an alternative sense of place for sites of natural heritage across the world.

The influence of the Earth Summit and this alternative engagement with natural heritage can be seen in the legislation which has emerged from the 1990s onwards. This has served to ameliorate the divide between the human and the non-human within the 'discourse of protection'. Whilst distinct boundaries remain and the non-human elements within this relationship are not accorded a role beyond being the object of humanity's engagement, places of natural heritage were framed as offering a site of cooperation within human society and with the environment. In Europe, this is reflected in the directive issued by the European Economic Community regarding the conservation of natural habitats and of wild fauna and flora in 1992 (known as the Habitats Directive). During an era when European integration was ratified with the Maastricht Treaty of 1992, the directive formed a central pillar of the European agenda for nature conservation. The ruling focused on preservation of national parks, nature reserves and sites of special interest as well as everyday spaces where the environmental legacy had shaped a particular locale. The rationale for the directive was based on the need to recognise and ensure the continuation of the continent's 'shared' natural heritage and to foster a sense of sustainable development within member states (Dodds, Strauss and Strong 2012). This gave a clear direction as to how Europe's 'natural heritage' should be engaged with to ensure a stable future for the continent. A sense of place is formed within this legislation where natural heritage becomes more than an area to preserve in its entirety but a space of political witnessing as new identities and values are constructed. The directive states:

> Whereas, in the European territory of the Member States, natural habitats are continuing to deteriorate and an increasing number of wild species are seriously threatened; whereas given that the threatened habitats and species form part of the Community's natural heritage and the threats to them are often of a transboundary nature, it is necessary to take measures at Community level in order to conserve them (Council Directive 92/43/EEC 1992).

The 1992 directive also set the foundation for the Natura 2000 project that linked various sites of natural heritage across Europe. Networks of protected sites, some crossing national boundaries, were formed through this initiative which saw the development of a tangible 'shared heritage' in Europe (Natura 2000). This relationship between the human and the non-human within European legislation was confirmed with the formation of the European Landscape Convention in 2000 (Voghera 2011). This treaty, also known as the Florence Convention, sets out how all sites of natural heritage should be managed and engaged with across Europe. Significantly, the convention made no distinction between the types of areas under its provisions; the concept of natural heritage was applied broadly to any form of landscape within member European states (see Jones and Stenseke 2011). What was important in this agreement was the recognition of these places as a 'common resource' for European society. The continuation of this theme within the legislation ensured that conservation and protection constituted an act of political witnessing. However, the convention also provided a further basis for a new relationship with natural heritage which formed a sense of an inclusive society where the human and non-human interact and engage. What emerged with the convention was an act of political witnessing, a directive towards modernity but also a recognition that 'the social' could encompass more than just people:

> Aware that the landscape contributes to the formation of local cultures and that it is a basic component of the European natural and cultural heritage, contributing to human well-being and consolidation of the European identity (European Landscape Convention, 2000).

It was this notion of working with the environment and the perception of natural heritage as part of a wider cultural and social system that highlighted a shift in the way in which sites and locales were regarded. The legislation set in place to protect and preserve areas of natural heritage, whether designated national parks or sites of special interest, has ensured that these spaces are viewed and experienced in particular ways. The 'discourse of protection' which envelops these locales defines a particular sense of place which is excluded from human engagement and activity. Whilst the need to preserve and conserve is not doubted, the way in which this legislation perpetuates the divide between nature and culture is significant.

Sites of natural heritage have been regarded as distinct and beyond the realm of human engagement. Indeed, evidence or potential of human engagement is strictly prohibited within some conservation schemes. However, over the course of the last few decades, *in-situ* areas of natural heritage have been redefined through new schemes of environmental management. These national and inter-national treaties have attempted to redefine the relationship between the human and the non-human in order to achieve objectives of 'sustainable development'. This movement has been undertaken as a response to the threat posed to human habitats by environmental change. The alteration has been significant as it has required legislation to reframe these places of natural heritage as sites of coopera-tion or harmony with the environment rather than spaces of non-engagement. Within these new frameworks, sites are reimagined as places where human engage-ment with the natural landscape constitutes a productive relationship. There is now a means of regarding society and nature as involved in the same system. However, whilst this has ensured the focus of legislation is built around cooperation with natural heritage, it still maintains the barriers between human and non-humans. Within these directives and international agreements, the heritage of nature is regarded as something that is acted on rather than as a participant within society. As such, whilst current legislation ensures sites are protected and part of the lives of individuals, groups and communities, it continues the same sense of removal and distancing this heritage from human groups. To regard sites of natural heritage as places which require preservation away from the actions of human society is to misunderstand their key role in the formation of the modern world.

Locales, spaces and landscapes which are regarded as possessing an environ-mental legacy can be assessed as places where society is reshaped and where the future is created. Rather than investing these sites with an assumed 'historic' status that demands that they should be protected from the shifting concerns of human society, they can be reimagined as areas that are intimately entwined with those fluid notions of power, place and identity within the contemporary world. In essence, these sites of natural heritage are where societies are formed and where individuals, groups and communities acquire the status of modernity. This can be observed in the way in which European legislation from the late 1970s onwards framed sites of natural heritage as significant, not just because of their history and environmental significance, but because of their ability to define a shared European identity. It is within these areas that a common regard for environmental protection and collective vision for the future could be developed. Therefore, natural heritage both affirmed the values of the modern nation state and supported the project of greater European integration. This ensures that the legal frameworks protecting sites serve as a means by which an act of political witnessing can take place. To look upon these places is to affirm the ideals of modernity. Nevertheless, this is a vision of the modern age where it is only the human element which is allowed to participate within society. As defined by the legislation that represents them, places of natural heritage reassert the divide between nature and culture that has dominated western politics, culture and ideology since the advent of the Enlightenment and industrialisation (Latour 1993). If such notions of modernity are not universal but indeed the result of socio-economic circumstance, then they can be reworked. If the arbitrary divisions between nature and culture are a product of the modern era then they can be re-aligned to reshape society.

What can emerge from this consideration is a legislative format that recognises the 'democracy of things' (Latour 2005). This alteration can transform the sense of place present within sites of natural heritage from one born out of the 'discourse of protection', where the desire to preserve an essential quality of the locale is preeminent, to a 'discourse of democracy', where non-human elements within this relationship are brought into society. This is not a process of politicising objects and features of nature that are inherently apolitical; it is a means by which natural heritage, so frequently placed within a framework of political witnessing, is allowed to be represented. In effect, the constitutive elements of natural heritage, flora, fauna, geology and hydrology, become 'matters of concern' rather than fixed points within society (Latour 2004a). 'Matters of concern' constitute 'tangled beings' with no established sense of meaning apart from a demand that they are placed as central in debates and constituted by a social dialogue which defines and redefines their place within the world (Latour 2004b). The non-human parts of natural heritage can thereby become part of society through a legislative framework that establishes a sense that these places serve as formative points in our modern world. These sites are not where the past is preserved, but where society can be redefined for the future. It is within these places of natural heritage that an act of social witnessing can occur which ensures a democratic engagement with the elements of the environment. The consideration of how a sense of place is formed with natural heritage at sites, locales, areas and landscapes can create a new relationship with the environment and redefine society.

Representing places and spaces of natural heritage

The sites of natural heritage across Europe and North America are also defined by their representation within the media and public sphere. The sense of place that is evoked within these areas is constructed by the discourses, imagery and associations used by websites, newspapers, tourist initiatives and guidebooks to discuss areas or locales where an environmental legacy can be witnessed. Whilst these frequently convey the accessibility and engagement with sites of natural heritage, there remains a firm sense of separation between the human and the non-human. Individuals and communities may well be encouraged to use these spaces for a range of activities but there remains an aspect of this representation that reinforces a divide between culture and nature. This is most frequently manifested in two forms within the various media that portray areas of natural heritage: the assertion of a chronological distance between a distant past and the present and the characterisation of areas as fragile and endangered by human engagement. It is this mode of representation that creates a sense of place which ironically dislocates humanity from the environment whilst seeking to establish a greater sense of connection. Whilst the great age of landscapes, geological formations or water courses is not challenged and the danger of over-exploitation of natural resources is apparent, it does not entail the exclusion of natural heritage from the processes of society. A democracy of things requires the participation of these areas and sites as 'matters of concern'. As such, the sense of place that could be formed through the representation of these places can establish a connection which accounts for the connections between the human and the non-human elements. Natural heritage can serve not as a place of retreat, away from the trappings of contemporary society, but as a site of change and reinvention.

In Britain, sites of natural heritage are frequently cast as demonstrations of ancient geological or climatic upheaval whilst also essential aspects of identity (see Crang and Tolia-Kelly 2010). The 15 National Parks serve as sites where long-lived 'native' flora and fauna can be preserved and where the historic natural environment that formed these landscapes can be maintained (see National Parks 2016):

- Brecon Beacons (Wales)
- Broads (England)
- Cairngorms (Scotland)
- Dartmoor (England)
- Exmoor (England)
- Lake District (England)
- Loch Lomond and the Trossachs (Scotland)
- New Forest (England)
- Northumberland (England)
- North York Moors (England)
- Peak District (England)
- Pembrokeshire Coast (Wales)
- Snowdonia (Wales)
- South Downs (England)
- Yorkshire Dales (England)

In each area, visitors are encouraged to observe the natural processes that formed the areas over millions of years as part of a distinctive narrative of identity (see Storey 2010). Within the official website of each of these parks is a 'discourse of origins' which establishes the values and perspectives of these sites as places of moral witnessing. This rhetorical device serves to establish a character for the area that is firmly based within the ancient geological processes that moulded the landscape. Whilst human interaction with these areas is acknowledged and the manner in which culture has shaped the spaces in both tangible and intangible ways is part of this narrative, the basis of the park is asserted as the natural events which physically shaped the locale. In effect, through this mode of representation, a sense of place that is moulded from such accounts provides a means of framing the site as a location absent of human engagement in its initial form. The places of natural heritage are thereby rendered as separate in their genesis from their definition through various socio-cultural, legal, political and economic frameworks. It is within these sites that we can witness the presence of nature in its original form with the incursion of culture in the form of human settlement a later addition (Catsadorakis 2007). The 'discourse of origins' gives a sense of character and a sense of place to these national parks, but it serves to place the human and the non-human as different plot devices within the same narrative. For example, the Brecon Beacons are described by the National Parks as a fusion of human and natural processes, but it is the latter which is the literal and figurative bedrock of the park:

> Like much of Wales, our landscapes are truly ancient, shaped by the Ice Age. The iconic northern scarp was deeply incised by glaciers and Llyn Cwm Llwch . . . the best preserved glacial lake in South Wales. There are also some well preserved glacial screes and moraines. A number of old quarry sites along the northern flanks of the Beacons contain fossil ferns. Nearly eight millennia of human activity have also moulded our landscapes . . . (Brecon Beacons 2016).

A similar mode of representation can also be found within the description of the Northumberland National Park, where the history of glaciation provides the origin point to the physical appearance of the landscape:

> The past two million years have been dominated by a succession of Ice Ages. Throughout much of the last Ice Age, large areas of Northumberland lay beneath fast-moving parts of the British ice-sheet. These ice streams moulded and literally 'streamlined' the land. The Cheviot Massif hills deflected much of the streaming ice around it to the north and south. Behind the Cheviot Hills lay an area of slower-moving ice. This difference in speeds gives the landscape its different character (Northumberland National Park 2016).

The timescales involved in the creation of these landscapes are vast, with entire epochs used to describe how the physical appearance of a national park has altered. For example, the South Downs National Park is presented as a place of ancient land and climatic formation, emphasising the perspective needed to grasp how such an environment has been formed:

> Imagine the South East as a shallow tropical sea – between 75 and 90 million years ago that is exactly what covered this part of the UK. The distinctive chalk ridge of the South Downs often described as the spine of the South Downs National Park was formed at this time after layer after layer of marine deposits were laid down. These deposits once formed a huge dome of chalk

stretching across to Surrey and North Kent. Weathering and erosion during the last ice age sculpted the landscape into its valleys, distinct hilltops and ridges (South Downs National Park 2016).

The 'discourse of origins' is also replicated within the official introductions to the national parks within Britain. These accounts place considerable focus on the 'natural history' of landscapes as the basis of the encounter with these sites. These introductions, produced in association with the National Parks authority, use geology and climate as points in a narrative that gives character to the site but also further alienates the human from the non-human (Marsh 2001; Speakman 2001; Tully 2002). Within the last decade, there has been a movement to ensure the presence of human groups and communities is acknowledged as part of these places of natural heritage. This has arisen out of a need to drive revenues forward in towns and villages within national parks which have been traditionally under-funded (after Catsadorakis 2007). Whilst humans are thereby asserted into the places of natural heritage, the division between nature and culture is still present. These tourist initiatives are supported by the British Government, who set out an eight-point plan for using the national parks as emblems of sustainable develop-ment in 2016 (DEFRA 2016). Significantly, this was phrased around the need to guide and develop the future. In essence, it defines national parks as arenas where the colonisation of modernity takes place, where the values and ideals that have shaped our society are replicated and preserved:

> This plan sets out our ambition to put National Parks at the heart of the way we think about the environment and how we manage it for future generations. We want as many young people as possible to learn about and experience the natural environment. National Parks are a great way in: inspiring environ-ments that can be lifelong sources of wellbeing, identity, adventure and pride (DEFRA 2016).

The framing of the national park as the longue durée scale might appear to ensure the inability to reconcile people with the sites of natural heritage as such distances are observed as conceptually unbridgeable (Bauer 2009). However, it is not the passing of millennia that casts divisions but the way in which human and non-human are represented as interacting which replicates current value systems within society. The division between nature and culture in this respect fundamentally alters the relationship between humans and the environment (after Latour 1993). In the formation of a sense of place where the 'discourse of origins' dominates the framing of the sites as examples of 'nature reserved' from society, the manner in which a wider society can be formed to consider how humans live with non-humans is obscured.

Whilst the geological and climatological formation processes are correct, the basis of the park is firmly located in these origins of the region, not necessarily as the product of a variety of agendas and concerns that have arisen through the interaction between humans and natural heritage. National Parks are sites where humans and non-humans engage in a relationship which defines values and ideals within society. Certainly, there are points within the narratives provided by these sites which acknowledge the way in which national parks are constituted by the relationships that were formed between humans and the environment in the past, present and potentially the future. National Parks use this relationship to stress the need for continual engagement and vigilance to ensure the preservation of areas of natural heritage. This can be observed in the narrative of place provided by the Dartmoor National Park in southwest England:

> Dartmoor's natural environment is unique. The granite massif rises out of the Devon lowlands to form a dramatic landscape with a distinctive geology, flora and fauna. This canvas has been modified by thousands of years of human activity but in less intensive ways than elsewhere in lowland Britain, at least in recent centuries. Here there is still space for people and nature to live side by side (Dartmoor National Park, 2016).

Whilst human interaction with these sites is acknowledged, it inevitably constitutes a step beyond the initial creation and origin of these distinctly 'natural' places. This inevitably obscures the role of these national parks as constructions of the present and arenas where the values of modernity are defined and asserted. Rather than being 'ancient' sites, these are places that are incredibly modern. This is not just born out of their status as protected areas during the latter part of the twentieth century. It is developed through their ability to convey the attitudes and values of contemporary society. However, it is through the 'discourse of origins' that an act of political witnessing can be recognised; a means of viewing these places of natural heritage as a reiteration of the divisions between nature and culture. The absence of humans within some of the formative accounts of places may well reflect the accuracy of scientific landscape surveys but it does prevent the relationship between the human elements and the non-human elements being fully explored. This is also present within some of the sites of natural heritage beyond the national parks in Britain. Areas such as the Jurassic Coast along the Devon and Dorset shoreline in southern England utilise the chronological distance between humans and nature as part of their appeal to visitors (Figure 4.2). The entire area of over 150km was awarded World Heritage status in 2001 and this status was conferred through the site's representation as a place of ancient natural formation. It is the 'discourse of origins' which establishes the absence of humans and the significance of 'nature' as the basis of the locale's narrative:

> The Jurassic Coast is a Walk Through Time. With its rocks and fossils we can uncover detailed stories from Earth's ancient past. Through its landslides, cliffs and beaches we can learn about the natural processes that formed the coast and continue to shape the world today (Jurassic Coast, 2016).

The area is represented to visitors as an engagement with the ancient processes of the natural world which have shaped our past and present environment (Cochrane 2008). The history of the area's involvement in the development of geological

Figure 4.2 Jurassic rocks form the prominent grey cliff at the base of the hill, Golden Cap, Dorset, UK. © The Trustees of the Natural History Museum, London.

and palaeontological knowledge is represented but it is the formation of the area through millions of years of history which forms the basis of the site's representation (see Brunsden 2003). The presence of human groups within the region is considered as part of the protection and preservation of the fragile coastal landscape for the purposes of sustainable development (Howard and Pinder 2003). However, it is the discourse of origins that establishes the sense of place at the site, the formative moment at which visitors are required to engage with the area. This is an area that is formed from more than just the physical landscape; it is an area where human and non-human interests meet (Brunsden and Edmonds 2009). A landscape which is defined not by the millions of years of geological development but the continuing renegotiation of the 'matters of concern' within a society that encompasses multiple elements. The division between nature and culture forms a sense of place that obscures the relations and associations between natural heritage and human society. Time, place and identity are revealed in this sense of place as natural heritage becomes an element outside of society rather than a part of that society.

The geological formations and the vast timescale of the Jurassic Coast provide a means of reflection as to the way in which this environment is comprised of various human and non-human agendas. As such, it is at sites of natural heritage such as this that the future of society at large can be redefined. The legacy of rock formation within the region does not necessarily solely indicate the ancient processes that shaped the landscape of southern England but how this area has been formed from our own recent anxieties and concerns. In recent years, the significance of 'geoheritage' has been used to address some of these issues (after Hallam 1983; O'Halloran et al. 1994; Joyce 1994). A range of scholars have drawn attention to the potential economic, political and social matters tied to the recognition of this part of the global environmental legacy (see Pena dos Reis and Helena Henriques 2009). Geoheritage and geotourism have both developed as a means of utilising geology as a resource and for demonstrating humanity's interaction with rocks and minerals of the Earth. Whether through mining, building, leisure or decoration, the role of this natural heritage within society has been made prominent through a variety of initiatives and research projects (Brocx and Semeniuk 2007). This has fostered a concern for 'geoparks', areas of significant geological formations which are protected and managed as an educational and economic resource. Both UNESCO and the European Union sponsored schemes from the 1990s onwards to develop geoparks (McKeever and Zouros 2005). In June 1991, UNESCO sponsored the First International Symposium on the Conservation of our Geological Heritage in Digne-les-Bains, southern France (Société Géologique de France 1994). Attended by geologists from across the world, the meeting gave rise to a joint statement entitled 'International declaration of the rights of the memory of the Earth'. Within this document, an increased regard and reverence for the heritage of the Earth was demanded as an alternative vision of the relationship between humans and the environment was promoted.

> We have always been aware of the need to preserve our memories – i.e. our cultural heritage. Now the time has come to protect our natural heritage, the environment. The past of the Earth is no less important than that of human beings. Now it is time for us to learn to protect, and by doing so, to learn about the past of the Earth, to read this book written before our advent: that is our geological heritage (Société Géologique de France 1991).

It is this approach to geoheritage which promotes a greater dialogue between humans and the environment. Within the declaration there is a strong assertion of natural heritage being a 'matter of concern' rather than a point of reflection or an object of conservation. In this manner, a focus on geoheritage enables the formation of moral witnesses as the 'Digne Declaration' focused on the issue of responsibility. This construction relies heavily on the role of humanity as 'gatekeepers' or managers of the environment rather than participants in society with other non-human agents of natural heritage:

> We and the Earth share our common heritage. We and governments are but the custodians of this heritage. Each and every human being should understand that the slightest depredation mutilates, destroys and leads to irreversible losses. Any form of development should respect the singularity of this heritage (Société Géologique de France 1991).

Therefore, whilst notions of geoheritage have provided greater concentration on how this resource is managed between governments, groups and individuals, it still reflects the wider ideological divisions within society regarding the relations between nature and culture. Within this scheme, the value of geoheritage is as an object of assessment, not as a 'matter of concern' within a broad society.

Across Europe, geoparks have been used to concentrate society's attention on the significance of the geological heritage and current environmental concerns. Whilst the sites have proliferated under European Union schemes of protection, the locales are still structured by a sense of place that can remove a sense of human engagement as 'nature' is protected and removed from society. The Massif des Bauges Geopark in central southern France is an example of this process as a vision of a 'pristine' area is presented to wider society as an escape from the modern word. As such, it is the 'discourse of origins' which frames this sense of place:

> The Bauges is mainly composed of sedimentary rocks formed under the sea between 180 and 66 million years ago. These rocks, folded during the form-ation of the Alps about 20 million years ago, have long been shaped by power-ful glaciers over 24,000 years which once extended down to Lyon (Géopark des Bauges 2016).

By 2016, just under 70 geoparks had been developed across the European Union and over 100 through UNESCO across the world. Within Europe, each geopark represented a unique era in the geological formation of the continent. These sites have raised the profile of geodiversity and as part of a European initiative they have also developed sustainability plans to provide for local communities (EGN 2000). Geoparks provide a site where nature and culture can interact and engage. This can be seen with the development of the Gea Norvegica Geopark in southern Norway which became a part of the European Geopark group in 2006. This site details the history of its geological formation from 1500 million years ago to the present, emphasising the way in which powerful geological forces have shaped the area and as well as the wider Norwegian landscape. The direct connection between contemporary society and the events of millions of years ago is achieved through detailing the range of areas of the geopark which demonstrate the different processes that formed the character of the place. However, across these sites is the connecting history of successive Ice Ages which have given rise to the particular physical character of the region. This natural heritage is offered to Norwegian society as a point of shared reflection, 'where the old Scandinavian geology meets the younger geology of continental Europe' (Gea Norvegica Geopark, 2016). The geoparks of Europe do provide an alternative engagement with the environmental legacy of the Earth but it is still one that is reliant on the permanence of geology rather than the recognition of its interpellation (after Patel 2012).

The role of geological formations as a part of natural heritage can be further explored with the example of the Giant's Causeway (Clochán an Aifir) in County Antrim, Northern Ireland. The site, which was formed through basaltic volcanism during the Palaeocene (60 million years ago), consists of a series of largely hexago-nal columns by the coastline which were formed as the lava slowly cooled. The area was granted World Heritage status by UNESCO in 1986 and is operated by the charitable organisation, the National Trust. The Giant's Causeway has been used as a site of 'geoheritage' as a means to inform and guide society on the processes of the Earth's formation over millions of years (Crawford and Black 2012). The area is defined by the Natural Trust as a place of both geological significance and as the place of myths, legends and discoveries which have shaped the site into a unique place:

> Flanked by the wild North Atlantic Ocean and a landscape of dramatic cliffs, for centuries the Giant's Causeway has inspired artists, stirred scientific debate and captured the imagination of all who see it (The Giant's Causeway, 2016).

The consideration of how society has interacted and engaged with the Giant's Causeway does alter the sense of place within the site. The entire area is presented as space of interpretation, where successive generations have brought new meanings and ideas onto the natural landscape (see Groves 2016). However, the 'discourse of origins' is still present within this perspective as 'nature' forms the seemingly neutral bedrock from which this narrative emerges. Whilst the area is revealed as a place of meaning-making, it is a process of creation without the 'matter of concern' itself. The imposition of interpretation does not constitute a democratic engagement with the environment and as such the potential for dialogue between the human and the non-human is obscured. Nature as witnessed at the site is always removed from the cultural. Of course, the timescales involved in the geological formation of the basalt columns could be regarded as a barrier to engagement but this is an issue of place as well as chronology. A sense of place that connects contemporary concerns with the Giant's Causeway as a part of a wider society can be altered by removing the distinction between nature and culture. Whilst the focus on science and folklore around the area demonstrates the construction of the place between humans and non-humans, the physical environmental legacy forms the backdrop or a mode of inspiration for successive generations of visitors (see Doughty 2008). Indeed, guides to the geological formation frequently assert the division between nature and culture, with the latter formed as a response to the site rather than constituting the site itself:

> For centuries countless visitors have marvelled at the majesty and mystery of the Giant's Causeway. At the heart of one of Europe's most magnificent coastlines its unique rock formations have, for millions of years, stood as a natural rampart against the unbridled ferocity of Atlantic storms. The rugged symmetry of the columns never fails to intrigue and inspire our visitors. To stroll on the Giant's Causeway is to voyage back in time (Giant's Causeway Official Guide, 2016).

The apparent neutrality of this site of nature as a place where people have created ideas about their world has been an important part of the development of heritage tourism within Northern Ireland (McManus 2009). Natural history, shorn of sectarian associations, would appear to provide a means of discussing culture and identity without the connections to the political and religious divides within the country (Graham 2003). Whilst this addresses existing tensions within human society, it does rely upon the strict division between nature and culture and the potential obstruction of a wider notion of society. To an extent, this reveals the way in which natural heritage resources are the object of varying demands and concerns from a range of interest groups.

This issue was addressed in June 2012 by 'geoheritage' interest groups in Scotland, including the organisational body Scottish Natural Heritage and the Scottish Government. These groups oversaw the production of Scotland's Geodiversity Charter. This document outlined the importance of Scotland's geological heritage to the nation for the development of the economy, the environment and the sustainability of society. The charter contained a series of principles for local authorities and stakeholders to develop their plans for geotourism whilst conserving the environment for future generations (Scotland's Geodiversity Charter 2012). The programme offered by the charter is innovative in its proposal that a greater amount of focus is required to ensure society works with the rich diversity of Scotland's geoheritage. As such, the country's extensive geological heritage, which records the movement of tectonic plates, the formation of mountain ranges, the effects of volcanism and climatic changes alongside the processes of erosion caused by weathering, becomes 'matters of concern' for all members of society (Scotland's Geodiversity Charter 2012: 2). However, despite the realisation of how this heritage contributes and shapes contemporary society, the 'discourse of origins' is still evoked to characterise sites of natural heritage as rooted and stable pieces of nature which evoke cultural responses:

> Scotland has world-class geodiversity – the variety of rocks, landforms, sediments, soils and the natural processes which form and alter them. Our geodiversity is vital as the foundation of life, providing essential benefits for society through its profound influence on landscape, habitats and species, the economy, historical and cultural heritage, education, health and well-being (Scotland's Geodiversity Charter 2012: 2).

The separation of culture and nature in this regard reiterates the relationship within western society that has characterised associations towards this environmental heritage. Although the Charter cites a progressive concept of working with geodiversity, the role of geological formations as being the 'foundation of life' obscures the participation of this element within society. Through its promotion as a physical and ideological 'bedrock', the profile of geoheritage is raised but its status as 'nature' is reasserted. This removes it from consideration as a part of society which encompasses humans and non-humans who are all constituted as 'matters of concern'. This entails that the highly important sites of natural heritage across Scotland, such as Fingal's Cave on the island of Staffa in the Inner Hebrides or St Kilda in the Outer Hebrides, which were formed alongside the Giant's Causeway 60 million years ago during a period of volcanic activity that created the Atlantic Ocean, can serve as evidence of past activity, not present participation.

St Kilda (Hiort) is an example of this process as its designation in 1986 as the only site provided with World Heritage status in Scotland due to its natural significance makes it one of the most valuable sites of 'geoheritage' in the country. The island, which has not been inhabited by humans since the 1930s, places emphasis on its status as a preserve, maintaining its unique geological character. Alongside this concern, the island's ornithological significance as a breeding site for seabirds is also an issue of engagement for The National Trust for Scotland who manage St Kilda and the wider environment (The National Trust for Scotland 2016). In recent years, the human settlement of the site has also been part of the conservation processes, with the homes abandoned by the islanders conserved for modern-day visitors (Fleming 2000). However, these areas are kept separate within the discourses that represent the sites; geology, nature and humanity are distinguished, as such divisions do not necessarily reflect the development of the site and the interactions between the geology and the environment that created and defined it. In part, this is result of the process of acquiring World Heritage status, which requires detailing the significance of areas for classification purposes (see Pocock 1997). Nevertheless, in 2005 St Kilda was awarded the distinction of dual World Heritage recognition for both its cultural and natural significance which, whilst emphasising the various processes which shaped the island, does still assert a division between humanity and nature:

> Already acknowledged for its magnificent physical beauty and its biological character, St Kilda has now been inscribed as a cultural record of a lost crofting community that once lived on what has been described as 'the edge of the world'. The remoteness of the islands – 64 kilometres west of the Outer Hebrides, off the west coast of Scotland – and the limited human interference over 5 millennia means it represents a highly authentic example of a way of life, now lost (The National Trust for Scotland 2016).

In a society where human and non-human are matters of equal concern not matters of distinction, such a division between the environmental character of the island and its inhabitation would appear irrelevant. Through a reassertion of that separation, the past human settlement of St. Kilda is regarded as part of a natural scenery whilst present human engagement with the site represents an incursion onto nature. As such, what structures this understanding of place beyond divisions of nature and culture is that of time. Notions and experiences of time are used to frame the experience of this site as part of a process of identity formation (after Jackson 1994). Areas of geological heritage are presented with their vast timescales of millions of years as a point of observation; the antiquity of the place is part of its separation from the 'normal' realm of things (after McGrath and Jebb 2015). As such, these places construct visitors as witnesses to the grand geological ages which shaped the planet. It is this long-term perspective which can both alienate and integrate the individual, community or society that connects to this natural heritage. The confrontation with a timespan which stretches into the millions of years can serve to disconnect engagement between humans and non-humans, where antiquity alone qualifies sites with the status of a preserved and protected space. In essence, this emphasises how places of natural heritage are formed not solely through geomorphic processes but also by the notions of endurance or perdurance within human society (see Lewis 1986). To distinguish a geological formation for its chronology would appear to remove it from engagement with wider contexts and in effect value it as an exhibition of the conceptual or scientific tools of measurement used to establish its age. This can be observed with the site of Siccar Point, Berwickshire, where the noted geologist James Hutton established his theory of unconformity in the 1780s. This area is defined as 'site of special scientific interest' but its significance is dependent on the discovery of its age. Siccar Point is classed on this basis as an object of reflection rather than a matter of concern:

> Siccar Point presents an impressive unconformity, with pale, horizontal Upper Old Red Sandstone laid down in the Devonian age (405–355 million years ago) lying on top of darker, older, vertical rocks laid down during the Silurian age (435–405 million years ago). The base of the younger rocks cuts across the bedding of the older rocks, showing that in the time interval between their formation, the older rocks were tilted and eroded (Scottish Natural Heritage 2010).

It is the 'discourse of origins' which is reiterated in this assessment as the site serves as a means to witness the Earth's antiquity and humanity's ability to measure that antiquity. As such, Siccar Point is presented as a place of endurance, a fixed point in the landscape that has persisted throughout the millions of years into the present. However, the site can also be presented as a place of perdurance, where the aspects of the natural environment have no fixed meaning but represent a variety of characters and definitions over time and space. This can serve as a means of reconciling nature and culture and provide a means by which this heritage can be part of society, not excluded from engagement. This can develop sites and a sense of place which can see natural heritage emerging as a 'matter of concern' (see Gordon and Baker 2016).

Being in places of natural heritage

Where geoheritage has been used as part of the social world, as a means of developing a common sense of place, the connections between humans and non-humans are at the forefront of this process. This can be seen at the site of Dan-yr-Ogof, part of the National Show Caves Centre in South Wales. This extensive, 17 km cave network has been shaped from the initial Carboniferous Limestone which formed over three hundred million years ago, through successive Ice Ages from two million years ago to the present day where the action of water still carves through the rock. The caverns which have formed through the process have become a prominent tourist site since their discovery in 1912 and were voted the finest 'natural wonder' in Britain by public polls in 2005 (Anon 2005). The site is not framed by this history and constitutes a sense of place not through the demonstration of its antiquity but of its continual relevance. The presentation of Dan-yr-Ogof is arranged as a process of discovery for the visitor; cave systems, underground lakes and rock formations are presented as an ever-developing narrative rather than a fait accompli. Indeed, the National Show Caves Centre (2016) encourages this response with its visitors by asking, 'who knows what might be discovered tomorrow?' As a 'show cave', Dan-yr-Ogof is part of a wider network of tourist sites across Britain that utilise natural heritage to promote tourist ventures which emphasise experience and engagement. These initiatives regard the age and formation of the caverns they represent but they seek to assert a different engagement with this environment. From Wookey Hole in Somerset, southwest England, to White Scar Cave in the Yorkshire Dales, these are tourist sites which regard the ancient formation processes that created these subterranean formations as a part of contemporary society. Whilst all commercial operations, the 'show caves' can also be linked through the emphasis these sites place on the experience, relevance and immediacy of this natural heritage, which is frequently cast as a process of encounter. For example, Ingleborough Cave in Clapham, Yorkshire, explains to visitors:

> At the end of the tourist path the cave passage is seen disappearing into the distance. . . . Even after all this time, explorations in the far extremities of the system continue to unravel the secrets of this hidden world (Ingleborough Cave 2016).

A sense of place is created here which embeds visitors into the site; a process that enables them to become more than just witnesses to the passing of time within an area which has been cordoned off from the modern world. Rather, within these locales they can be witnesses to the process of time and nature as part of that place (after Gordon 2012). This reflects how concepts of perdurance, performance and poetics are significant in the formation of a sense of place with natural heritage (after Baker and Gordon 2012). This is associated with the development of 'geopoetics' which rejects the enduring model of time and place within environmental sites. Instead, geopoetics encourages the perception that different values, definitions and timescales can be present within locales and landscapes of natural heritage (after White 1994; 1996). It is this recognition which allows for the emergence of a 'discourse of experience' that creates a sense of place through an engagement with the sights, scenes and stories of a locale (Gillet 2009). In this manner, natural heritage is not a backdrop, a demonstration of antiquity nor the cause of an underlying character of a place, it is a 'matter of concern' which is constructed in relationship with wider society (after Westphal 2007). This mode of engagement can be regarded as an act of social witnessing. It requires the individual or group to encounter the environmental legacy as elements that humans exist with. As such, it is a clear expression of the democracy of things where the human and non-human interact and develop new social values on the basis of engagement. This occurs through a 'discourse of experience' where places are represented as part of the past, present and future. Whether through art, literature or performance, natural heritage becomes an integral part of society rather than a preserved space removed from wider contexts.

Such an endeavour can be observed with the creation of trails and installations within natural heritage sites which engage visitors with the experience of a creation of place and time rather than an imposition of the value of space (after Patullo 1997). In 2006, this particular approach was utilised for the Scottish Borders James Hutton Trail, which was launched as an initiative by the government body Scottish Natural Heritage alongside other interest groups, funding organisations and tourist boards. The extensive route was designed to guide visitors through the locations where the noted eighteenth century geologist observed changes within the Earth's structure alongside the sites which fostered and developed this knowledge. The trail originally consisted of sites near Jedburgh in the Scottish Borders. It included the farmhouse where Hutton lived in Coldingham, the marl-pit where Hutton extracted calcite-rich mud as a fertilizer, Siccar Point and finally a sculpture by the artist Max Nowell in Lothian Park which is entitled 'Hutton's Unconformity' and commemorates the geologist's observations at nearby Inchbonny (Browne and Floyd 2006). This trail was also supported by the development of sites in Edinburgh as well which brought visitors to encounter new places and old sites in different ways:

- James Hutton Memorial Garden, Edinburgh. On St. John's Hill in the Pleasance is a garden, originally built in 2001, where a display of boulders, each with veins of granite, is used to demonstrate Hutton's theory regarding the nature of igneous rocks but also the continuity and cyclical processes of geology.
- Salisbury Crags Sill, Edinburgh. This escarpment is approximately 300 million years old and was used by Hutton to demonstrate the volcanic activity that forced strata upwards. The presence of such a dynamic event within Edinburgh enables a reflection on place, time and humanity's relationship with the Earth.
- Hutton's Grave in Greyfriars Churchyard, Edinburgh. The grave of Hutton has been marked by a small plaque since the 1940s and as part of the trail it provides a reflection on how humans have engaged with their environment.
- Hutton Roof, National Museum Scotland, Edinburgh. On the terrace of the museum, which opened in 1998, the sculptor Andy Goldsworthy displayed four blocks of red sandstone from the Locharbriggs Quarry as a mediation on past, present and future. The display looks out onto the city as the stone, formed hundreds of millions of years ago from grains of sand and through erosion returning to that form, connects contemporary visitors with the processes that have shaped and still shape the planet.

(LaBRIGS 2006)

The 'discourse of experience' is drawn upon in these trails to emphasise the process of discovery and incorporation of natural history into the lives of those participating in contemporary society. Through a variety of forms and media, natural heritage becomes an element within society as a shared matter of concern through its inclusion within these heritage itineraries and walks. It enables an act

Figure 4.3 Salisbury Crags, Edinburgh.

of social witnessing as guides physically and conceptually develop associations between ancient geological events and everyday concerns. This relates to wider theories within the social sciences where the act of movement establishes structures of knowledge and experience (after Tuan 1977). Indeed, in some regards, the process of organised paths and trajectories in the landscape can be regarded as the locus of control as well as dissent within societies (after Sheets-Johnstone 1999). It is, therefore, through movement that the process of revelation for the individual and the wider community is formed (after Appadurai 1995). The 'heritage trail' which appears so ubiquitous within contemporary public history projects, which has its origins within the classical tours of the Italian peninsula in the eighteenth century, thus forms a key part in affirming norms and developing new relations within society.

> ... the rooting of the social in the actual ground of lived experience, where the earth we tread interfaces with the air we breathe. It is along this ground, and not in some ethereal realm of discursively constructed significance, over and above the material world, that lives are paced out in their mutual relations (Ingold and Vergunst 2008: 2).

It is through trails, guides and walks that a 'discourse of experience' is enacted and performed and provides a means for alternative experiences of natural heritage to be created. Whether organised by national organisations or created by local initiatives, the process of movement within these trails provides points of connection to a past, present and future.

The official trails produced by government agencies or national parks can provide a means of cementing social relations and divisions between nature and culture. Guides which emphasise the distinction and separation of environmental heritage from society as a means to establish the need for protection and conservation raise important concerns but can obscure the ways in which humans and non-humans build relationships. However, as demonstrations of the connections within a wide concept of society, these trails, guides and walks can serve to build a greater awareness of the role of natural heritage within social life. Various schemes across Europe, both regional and national, many of which have drawn funding from the European Union, have proliferated within the last ten years, resulting in walking or hiking trails in the national and natural parks of member states (see Calaforra and Fernández-Cortés 2006). These have provided local residents and tourists with a point of engagement with the environment, emphasising human participation in the world in the past and the present. For example, the Karst en Yesos de Sorbas, a nature reserve (paraje natural) in the southern Spanish region of Andalucía, has used its hiking trails to connect the area's modern history of mining and migration to its ancient geological formation. The area is rich in gypsum (yesos), a sulfate mineral which is comprised of calcium sulfate dehydrate; as well as a specialised flora, cave systems and a unique geology, a variety of human processes associated with the use of gypsum have shaped the area from the prehistoric era onwards. This co-existence is part of the heritage trail that ensures that humanity's engagement with the landscape is not 'cleaned' or 'removed' in order to preserve and restore a pristine wilderness. Rather, the trail connects the various elements of the area's development from millions of years ago to the present to enable a recognition of a shared sense of place (Sendero Señalizado Los Yesares 2016).

The use of trails or guides within national parks and nature reserves in France has also developed significantly since the beginning of the twenty-first century, with increasing numbers of national parks using this device to engage tourists (see Guiomar 2010). In the regional nature park in Vexin, north western France, a series of trails have been created to enable an appreciation of the natural heritage as well as the cultural heritage of this area of the country. The park was designated in 1995 and covers an area of approximately 710 km^2 centred around a large limestone plateau (after Darly 2010). Whilst supporting a rich natural heritage in the form of the native flora and fauna alongside its geoheritage, the park also possesses architectural and historical significance. From 2011, the park's management, under the local regional authorities and municipal governments, has developed over thirty trails throughout the protected area to establish a mode of engaging with visitors. Within these guides, visitors are provided with a route through the natural environment to observe the manner in which humans have interacted with their surroundings (PNR du Vexin 2016). Therefore, unbound by chronological frameworks and enabling a different appreciation of the relationship between individuals and communities alongside this natural heritage, the sense of place that is created within this area is entirely built towards the social (Brämer 2009). The focus here is within a 'discourse of experience', a mode of enabling people to regard the natural environment in a variety of ways through movement (see Brämer 2010). Since the 1990s, this has become a significant part of the tourist initiatives within German national parks (Vogt 2009). As well as being a commercial venture, the provision of hikes, trails and guides within German national parks has also been specifically linked with concerns for health, conservation and society (see Brämer 2007). This approach is present in the walks and tours available within the Eifel National Park, North Rhine-Westphalia, Germany. Eifel is a nature conservation area which covers over 100 km^2, including both natural and cultural heritage. A range of walks are provided on the site:

- Hiking with rangers
- Boat trips with rangers
- Walks accompanied by sign language
- Spiritual walks along the Creation Trail

(Nationalpark Eifel 2016a)

With a mixture of educational, cultural and social tours, the range of experiences within the area is extended. The provision of a Creation Trail was designed to enable visitors to the park to meditate upon the natural environment through spiritual texts and installations such as a labyrinth formed from small rocks (Nationalpark Eifel 2016b). Forming a connection between natural heritage and the various ways it can be constructed and understood, whether through a religious, scientific or social agenda, can enable the environment to become a 'matter of concern' within society. It enables an engagement with how humans interact with non-human elements within society and how this can establish a democracy of things. In this 'discourse of experience', guided trails provide a means by which new connections can be built within a broad concept of society.

Over the past two decades, routes defining natural heritage for both local residents and tourists alike have proliferated across Britain as a desire for local history has increased, de-industrialisation has left large areas of unused mines, quarries and factories, and a greater concern for environmental issues has also encouraged greater interest in sustainability and conservation. These trails have included walks through national parks and nature reserves as well as inner-city areas, as from the small-scale rambles to the extensive, several kilometre hikes, this is a process of engaging with natural heritage. For example, Kingly Vale in West Sussex, a National Nature Reserve (NNR) which is managed by Natural England, the public body responsible for protecting natural heritage, has a wealth of natural and cultural points of interest through the 1.6 km^2 site. From the chalk downland that supports a range of flora and fauna to the presence of ancient yew trees estimated to be at least 2,000 years old and some of the oldest living organisms in Britain, the site presents a range of interests to visitors. Guides have been produced to represent this unique landscape but they also serve to connect individuals, groups and communities using this space with all aspects of activity on this site. Both the human and the non-human are presented as part of the creation of this particular landscape as numbered posts along the site prompt visitors to reflect on the processes that have shaped the area. Indeed, the guide specifically lists both human and non-human aspects of the site side-by-side without hierarchy or distinction. This relationship is highlighted as key to understanding the area:

> Humans have been inextricably linked with this land from the earliest times and we cannot understand this landscape without appreciating their influence (Natural England 2013a: 30).

Similar guides for other NNRs produced by Natural England also reflect this mode of engagement and the use of the 'discourse of experience'. For example, Walberswick National Nature Reserve on the Suffolk coast introduces its guide with a recognition of the entwined relationship between humans and non-humans: a 'combination of the forces of nature and the hand of man have created the landscape we see at Walberswick today' (Natural England 2013b: 2). This use of direct experience with a site is a result of the organisation's commitment within the last ten years to increase public awareness of the environment within the country (Natural England 2008: 11). This is a programme which was encouraged by the British Government but has manifested itself into the basis of an alternative relationship between people and natural heritage. This has arisen as it is through moving through places and engaging in a range of times, concerns, subjects and connections that a social understanding of a place can be formed (after Miles 2013). In this manner, trails such as this reflect the 'discourse of experience' as they emphasise the process of encounter and exploration with the wider environment. Social witnesses are formed through these activities as they inscribe areas with a sense of place that allows for a recognition of past, present and future concerns through collaboration and cooperation with natural heritage.

In 2013, the Todmorden Moor Restoration Trust and West Yorkshire Geology Trust cooperated to develop the Todmorden Moor Geology and Heritage Trail as part of a regeneration programme for the former industrial area in the Calder Valley, West Yorkshire, England. The four-kilometre trail encompasses geological formations and prehistoric settlements alongside the expanse of mines and quarries that dotted the area to extract coal and sandstone. As such, the vast expanse of geological time is not isolated from the way in which humans have inhabited the area:

> These help us understand how the geology and man's exploitation of what they found have created the landscape that we see today. The panels also provide information on the fossils from more than 315 million years ago that can be found associated with the mine wastes (Todmorden Moor, 2013).

If a space where humans and non-humans can be part of a society is to be formed, then narratives and spaces where this interaction could potentially occur can be constructed through guides and trails. It is in this 'discourse of experience' that a sense of place is formed that connects individuals to the environment in a way which builds relationships between people and places. Through a process of engagement and discovery, the ancient landscapes and geological processes can be part of the modern world, not alienated by the chronological distance that can separate this past from the present and the future. The traditional distinction between nature and culture is, thereby, rendered insensible in considering how humans and non-humans have both acted to shape the natural heritage (after Latour 1993). Whilst trails and guides can serve to reinforce such distinctions, by cataloguing, defining and measuring sites for their age and formation processes without reference to their social context, they are also the site of revision and reinterpretation. The role of these routes is crucial in the development of a democracy of things where natural heritage becomes a 'matter of concern' within society. The 'discourse of experience' does not entail the prioritisation of human experience over all other elements. Rather, it prioritises the relationships and connections formed between humans and non-humans.

Conclusion

The formation of a sense of place through natural heritage has characterised the modern world. Indeed, it is within these sites, whether national parks, nature reserves or protected areas, that society achieves a status as truly modern. By separating 'nature', enacting legislation to preserve it in a state prior to human occupation, emphasising its distinction as a 'conserved space' over 'developed places' and ensuring its protected status to prevent alternative understandings of these sites, western nation states act in the most modern ways imaginable. In this act of modernity, the past is rendered distant whilst the present is mobilised to impact upon and shape the future (after Giddens 1990). The modern distinction between nature and culture has been the subject of various critiques in recent years as inherently limiting to how contemporary society could be organised (see Latour 2005). Sites of natural heritage are key in this process as they constitute the basis of this division but they can potentially serve as locales of resistance. Although legislation has succeeded in recognising a greater relationship between humans and nature, with both national and international agreements concurring on the basis of 'sustainable development', this is still an association framed by the divide which has been fostered in the modern era. The 'discourse of protection' inevitably ensures that natural places are regarded as apart from wider society, which sets a framework as to how they are perceived and engaged with. The representation of natural heritage as distinct due to its status as formed by some immense ancient process or event can also act to define places as separate from humanity.

Whether it is rock formations millions of years old or landscapes carved by successive Ice Ages before modern humans evolved, the 'discourse of origins' asserts the status of these places as beyond social comprehension. The emphasis on the formation of these places obscures how they have become part of a social, political, cultural and economic nexus which defines and redefines them throughout their history. It is this connection between humans and non-humans that can be emphasised to develop a new set of relations beyond the division made between culture and nature. In the 'discourse of experience', a sense of place is formed with the elements of natural heritage; as such, the environment ceases to be an object of contemplation but an arena in which people live and redefine society. It is in this manner that the trails, guides and walks that now appear to dominate the heritage industry become significant as they represent the potential for change. Within the process of moving within these spaces, individuals, groups and communities can establish links across time periods and subjects to see natural heritage as not just defined by one time but defined by the connections that are made with these places. Trails enable a reconsideration of space and time as they provide a means of reconfiguring the world and its comprehension. Essentially, these modes of access that are provided by public and private groups alike ensure the function of the social witness; the observer can look upon these places and recognise not the abstract agents of ancient formation but the real elements of change and process that shape past, present and future alike. The focus on connections provides the basis for an alternative vision of the social which includes all aspects of natural heritage.

5   Heritage, memory and natural history

Introduction

The representation of natural history within popular culture emphasises the presence and use of this heritage within contemporary society. Rather than being consigned to a meaningless corner of the past as nation states have raced forward into the industrial and technological age, natural heritage has remained central within western society as a mode of representation and as a vehicle for transformation. Therefore, even as modern society appears to be ever-increasingly removed from a natural environment, individuals and communities are intimately connected with this heritage as a means of reflection and demonstration. The popular culture of natural heritage can be examined in both tangible and intangible forms, as physical places, scenes, spectacles and language reflect how the environment is regarded within society (after Mitman 1999). These responses create a sense of place, identity and power as associations, references and direct uses of this past in the present establish social relations. In effect, political, moral and social witnesses are formed through this popular culture as the use and engagement with these media requires observation and testimony. The association with natural heritage, whether as a tourist, as an audience member or as a viewer, affirms the role of the witness by placing the individual in relation to the events they observe and requiring them to bear that burden of knowledge as to the effects on the world. In this manner, popular culture becomes more than just the reflection of consumerism and media representation, it serves as a means of expression (after Mosley 2006). Individuals, groups and communities utilise natural heritage within popular culture as a means of establishing value. By assessing this engagement, the study of natural heritage can be moved beyond the analysis of legislation or the structure of institutions to how these concepts are experienced and deployed within society. Through the examination of tourist initiatives, media representations, visual culture and language, the relationship between humans and non-humans can be regarded as a process of engagement and negotiation (after Olwig 2010). Therefore, it is through popular culture that an alternative assessment of society can be formed.

Television, film and theme parks provide frames through which natural heritage is comprehended (Mitman 1993). These are not static frames; instead, they are actively used by audiences to imagine and reimagine their relationship with the environment in the past and the present (after Wertsch 2002). Rather than assume that societies passively consume media images, the way in which these images prompt responses or serve as tools for alternative ideas can be examined. This has been a relatively understudied field of enquiry with natural history's role in popular culture alongside natural science having been largely avoided by scholars across the discipline (after Cooter and Pumfrey 1994). This neglect is in contrast to the development of environmental studies across the humanities. This work has demonstrated the ways in which rhetoric and discourse have defined the history and legacy of the natural world within society (see Killingsworth and Palmer 1992; Myerson and Rydin 1996). Scholars have also highlighted the ways in which the media representation of environmental concerns in news programmes, documentaries and campaign videos is a significant factor in disseminating values and agendas with regard to managing natural heritage (Anderson 1997; 2014; Lester 2010). When this is combined with the ever-growing area of environmental political studies and political ecology, despite the prominence of individuals' works, the general absence of a sustained assessment of how natural heritage functions within popular culture or cultural studies appears to be somewhat curious (see Schama 1995). Certainly the presumed 'natural' state of the environment might ensure that this heritage can appear beyond the act of interpellation. Despite this appearance, the representation of natural heritage within society is structured by the values, anxieties and agendas of those within that society (after Meister and Japp 2002; Parham 2016). This portrayal can be a platform for the act of the social, moral and political witness; therefore, it is through these mediums that norms can be asserted but revisions of society can be achieved.

Dinosaur parks: places of past and present

The use of statues and recreations of dinosaurs has been a feature of museums and exhibitions since the mid-nineteenth century when the display of ferocious animals delighted audiences. Whilst the provision of life-like dinosaur models remains a significant aspect of the work of natural history museums, since the 1950s tourist sites have been developed across the United States and Europe which house full-scale sculptures of these extinct creatures as an attraction for paying visitors. These parks or display grounds are far removed from the setting of the museum and the institutional mission to educate and inform. Whilst these areas may offer knowledge and information on the appearance and habitats of the dinosaurs, the purpose of these displays is to engage as part of a commercial venture. Nevertheless, this function does not prevent the dinosaur parks, which have proliferated since the 1990s, from constituting a highly important means of understanding natural heritage and considering the contemporary place of humanity within the world. Rather than being dismissed as a tawdry commercial enterprise or merely as children's entertainment, these locations actually demonstrate their value from such associations. They reflect a status of objectification and desire but also serve as a formative point in the relationships individuals and groups within society form with the environment. In effect, these places serve as points of social witnessing, where the relationship between humans and the wider natural heritage can be demonstrated. The arrangement and display of these parks can be considered a return to the use of emotional engagement through the creation of sublime sights in a fashion similar to the shows and performances of nature in the eighteenth and nineteenth centuries. However, contemporary dinosaur parks move beyond the assertion of one's own existence in the face of a long extinct monster which was induced by the solipsism of those historical saurian displays. What is evident within these arenas is a creation of a sense of place which encompasses a reflection on the social. It is not the extinction of these animals that is significant in these circumstances: it is their relevance for present day society (Mitchell 1998).

These parks developed from the 1930s after the awe-inspiring display of moving dinosaurs at the 1933 Chicago World's Fair. This exhibition, entitled 'The World a Million Years Ago', was sponsored by the company Sinclair Oil. The firm subsequently adopted a dinosaur logo of a Brontosaurus after the show proved highly-popular. This was accompanied by a launch of Brontosaurus models, signs and displays by gas stations in the United States which saw the emergence of competing scale model dinosaurs across the country. One of the first of these was the eight-metre dinosaur constructed outside of the Creston General Store in Creston, South Dakota. Further examples such as Dinosaur Gardens, Ossineke, Michigan, which featured an extensive collection of sculptured animals developed during the 1930s, whilst Dinosaur Park in Rapid City, South Dakota, was developed as a tourist attraction with large-scale concrete animals arranged in dramatic poses on a crest of a hill overlooking the city below (Figure 5.1). This can be considered to be the basis of the dinosaur parks that have been created through the latter half of the twentieth century in Europe, Canada and the United States. Designed as a commercial enterprise, these parks, gardens and displays have been an important tool in integrating natural history within wider popular culture. They have also been used as a device to challenge evolutionary science as part of a religious world view as dinosaur parks from the 1960s have frequently asserted a Creationist stance. Such displays misrepresent data and erroneously label evidence to support a vision of the Earth that was created by a deity. In effect, they perform the same role as the displays of the mid-nineteenth century, utilising the models to reflect a sense of self and superiority. This is accomplished through the medium in which this heritage is delivered. The model of the amusement park provides an alternative perspective on how natural history is incorporated into contemporary life. It is in this environ that heritage can be associated with a sense of being; whilst implicated within capitalist economics the attachment to leisure and pleasure entails a recognition of a social existence. Indeed, within the area of the dinosaur park, the Bergsonian notion of the élan vital can be observed alongside the formation of the social witness.

The contemporary tourist landscape within the United States is awash with dinosaur parks. Varying in scale, size and mission, these sites provide a confrontation with life-size replicas of the extinct animals. In this respect, they differ little from the museum exhibitions which may rely upon animatronic recreations or models as

Figure 5.1 Dinosaur Park, Rapid City, South Dakota. Carol M. Highsmith Archive, Library of Congress, LC-DIG-highsm- 04620.

part of a wider programme of accessible public displays. What is distinct about these sites is that, whilst they may provide scientifically accurate details about their displays, their arrangement is frequently based not on evolutionary taxonomy or chronological period but on the effective mode of engagement with the individual. These are places which are designed to elicit an emotional response as a touristic experience but also a social act as society reflects upon its own place within a far wider scheme. This is evident in the array of parks in the United States and Canada which operate under the 'Dinosaurs Alive!' label which is owned by the leisure group Cedar Fair Entertainment Company. Since 2011, nine parks have opened which are attached to existing theme parks or leisure facilities. The largest of these Dinosaurs Alive sites is at Kings Island theme park, Mason, Ohio, where over 60 animatronic dinosaurs are displayed in a trail over one kilometre in length. The scale of the venue and the volume of models gives the location the distinction of claiming to be the largest dinosaur park in the world. This particular site provides visitors with a range of activities and a series of scenes to encounter, all of which are limited not by time or space but encompass an engagement with the past, present and future.

> Long before coasters, dinosaurs ruled Kings Island. Now they're back and they haven't eaten in a very long time. Come face-to-face with the most animatronic dinosaurs on Earth at the only place big enough to hold them all. . . . Plus, find out what it's like to be a paleontologist at our dig site. From

the ferocious Tyrannosaurus rex to the soaring Pteranodon to the largest animatronic dinosaur in the world, Dinosaurs Alive! is the dawn of a new era of thrills. Dinosaurs Alive! – The biggest thing in 65 million years (Dinosaurs Alive! 2016a).

The site represents the creation of a distinct sense of place as the park attempts to forge a connection between contemporary visitors and animals extinct for millions of years. The scale and existence of the dinosaurs is emphasised in these displays as the park provides a space where artefacts are transformed into non-human objects. Animatronic dinosaurs are depicted in a number of different scenes to detail their life and engagement within the world. The park provides the following scenarios:

- Pack Attack – grazing herbivores are attacked by carnivorous dinosaurs;
- Flash Flood – the death of a group of dinosaurs drowning in a fast-flowing river is highlighted alongside the processes of deposition and fossilisation;
- Sauropods – the variations of dinosaur types are explained through a grazing scene;
- Predator Trap – a group of dinosaurs becomes stuck in a mud pit with other dinosaurs awaiting their fate so they feed off the carcasses;
- Kids Dig Site – a recreation of an excavation of a dinosaur skeleton immersed in sand is provided for children;
- T-Rex Fight – Triceratops are pursed by a group of Tyrannosaurus Rex;
- Fight to the Finish –fleeing dinosaurs escape from predation by a Tyrannosaurus Rex.

Within each of these dramatic reconstructions of ancient life visitors are moved beyond the human to consider an environment where their place in nature is not dominant (after Pearson 2007). The sublime sight of herds of dinosaurs roaming the hills of the park constitutes a movement away from the self to contemplate the social and the élan vital as they offer a point of reflection on human and non-human. Removed from the dominance over life on Earth and placed within an environmental heritage where the physical life of extinct creatures is clearly portrayed, individuals, groups and communities become social witnesses. Such recognition is not born out of the robotic mechanisms which enabled the simulation of life. Rather, this is achieved through the placing of humans within this natural heritage as an ongoing and engaging concern. The sense of place that is created at such sites therefore represents an incredibly complex construction; contemporary human society is required to observe the passage of a now extinct life of non-human society in the same space we inhabit (after Kohn 2013). Rather than constituting a paradox of time and space, this constitutes an engaging form of transcending the confines of 'the human' to consider 'the social'.

The other Dinosaurs Alive! parks within the United States and Canada follow a similar pattern of engagement. The models and scenes for these sites are provided by a private company, Dinosaurs Unearthed. This group has been operating travelling exhibitions and displays of saurian animatronics since 2005 and utilises a range of technological skills to ensure a 'life-like' appearance and movement within their recreations (Dinosaurs Unearthed 2016). The locations in Virginia, Minnesota, Ontario and Pennsylvania, whilst differing in size and the number of specimens displayed, arrange the exhibits on a path where the lives, deaths and discoveries of dinosaurs are encountered as a visitor experience. The connection with scientific accuracy is part of these trails through the dinosaur parks but this is presented as a narrative entwined with the excitement of engagement. In King's Dominion, Doswell, Virginia, the Dinosaurs Alive! site provides scenes where the creatures graze, defend their young or fall victim to the environment, as a replication of fossil evidence alongside the human experience of the present:

> This dinosaur exhibit presents scenes and stories based on real fossil evidence, such as the predator trap at Cleveland-Lloyd in Utah and the Pachyrhinosaurus flash flood episode from Pipestone Creek in Alberta, Canada. Other key scenes depicting behaviors and diversity include an attack by a pack of Deinonychus on a grazing Hadrosaurs, and an adult and sub-adult Tyrannosaurus Rex stalking a Triceratops! (Dinosaurs Alive! 2016b).

Significantly, these sites are intended to provide more than just points of individual reflection as the passing of time between past and present is considered when facing the models of long extinct creatures. Indeed, these parks attempt to create a social environment where contemporary visitors can consider associations beyond the present and beyond an anthropocentric system of thought. It is within these parks that the sublime and the symbiotic are called forth as a means of witnessing the presence of others:

> The opportunity to observe the size of these massive hadrosaurs and learn that they lived in herds numbering in the hundreds and thousands while sharing their ecological environment with even larger predators, becomes a moment of awe and understanding for guests of all ages of a world long since extinct (Dinosaurs Alive! 2016c).

This is a theme which is shared by other dinosaur parks across the United States and Canada. The three sites in Cave City (Kentucky), Glen Rose (Texas) and Plant City (Florida), which operate as part of the 'Dinosaur World' group, all feature life-size dinosaur models, trails through replicated scenes and excavation zones for children to recreate the activities of a palaeontological dig (Dinosaur World 2016a). The park in Plant City opened first in 1998 and features over a hundred models of dinosaurs and other extinct creatures within an area of approximately 0.08 km². This Floridian 'Dinosaur World' opens up a vista on humanity's engagement with natural heritage as it provides points of excitement and spectacle alongside a recognition of the past lives of these animals. Central to this process is the physical acknowledgement of this history as lived and experienced in the present:

> With over 20-acres to explore, you will find yourself wandering amongst some of the most feared predators ever to walk the planet. Dinosaur World offers a unique, natural setting featuring a lush garden setting and paved walkways where you will encounter a wide variety of dinosaurs representing many diverse and unusual species (Dinosaur World 2016a).

The focus on 'being' with the replica dinosaurs in a recreated setting where the variety of extinct life is presented replicates the definition of la durée forwarded by Bergson (1910: 90–91). The notion of existence in this manner is not posited purely in the present but connects to previous and potential other modes of existence as 'real time' is experienced apart from 'mechanical time' (Bergson 1910: 127). The arrangement of the dinosaur models as a process of encounter rather than as a demonstration of typology or chronological order alters the position of the past in relationship to the present and future. The exhibition of life within another chronological context and its participation in the here-and-now calls forth a new type of present:

> The past does not cause one present to pass without calling forth another, but itself neither passes nor comes forth. . . . We cannot say that it was. It no longer exists, it does exist, but it insists, it consists, it is. It insists with the former present, it consists with the new or present present. It is the in-itself of time as the final ground of the passage of time (Deleuze 1994: 82).

The multiplicity of time and space in this movement through the models potentially alters both the present and the future as alternative points of consciousness are formed through the engagement with this representation of natural history. The park is organised around a series of experiences, with visitors observing the processes of an archaeological dig and handling excavated fossils before observing the exhibits. The walk around the site encompasses an exhibition of herbivorous dinosaurs but particular focus is placed on the 'Carnivore Boardwalk'. On an elevated platform, scenes of dinosaurs stalking and devouring their prey are presented as an experience of fear and fascination. The confrontation with the state of nature in the past allows for a contemplation of humanity's role in the present. In observing an era where non-humans are in the ascendency as an event occurring in the present the status of humanity as the agents of definition and meaning in the world is challenged. In essence, it enables a recognition of the social environment as more than just the relations formed amongst humans. This mode of engagement is also evident in the park's use of animals that did once live alongside humans, as an expansion to include extinct Ice Age creatures has seen the presentation of Mammoth in a range of family groups and individual settings. The entrance to the display is a re-created cave with early hominid paintings reproduced along the walls. The guided trail through the mammoth herd presents data on the human use of these animals alongside their lifecycles and adaptation. As such, it offers a point of connection beyond the human perspective in the past but also the present as the associations with the natural environment are brought into consideration:

> Stroll with the Giants of the Ice Age! These woolly mammoths are fierce and large, appearing as they did during their time. Watch your step for ice on the ground and check out the educational signs to learn more about mammoths and their existence with man (Dinosaur World 1916b).

The insistence upon the presence and present status of this past provides a recognition of a social environment, altering the perspective of contemporary visitors to the park. This sense of conjunction is a feature across dinosaur parks in the United States and Canada. Indeed, sites demonstrate their relevancy and advertise their appeal through this association. It is through this connection that dinosaur parks can act as places where alternative visions of society are formed. For example, the 'Field Station Dinosaurs' site, which opened to visitors in 2012, initially in Secaucus, New Jersey, before relocating in 2016 to Overpeck County Park in Leonia, New Jersey, offers its visitors a vivid point of connection between the modern urban landscape which surrounds it and the models of extinct creatures. Indeed, its slogan reinforces this image, not as a contrast but as a means of giving a context to eras separated by millions of years: '9 minutes from Manhattan 90 million years back in time' (Field Station Dinosaurs 2016). The site is set within 0.08 km^2 of woodland and lawns with just over 30 model dinosaurs displayed alongside trails through the area. Common features of dinosaur parks such as a recreated palaeontological excavation are present, as are the recreated scenes of grazing, hunting and family groups. These displays are within a short distance from New York which presents a means of emphasising relevance and ontemporaneity:

> The Field Station is an oasis of natural wonder just minutes from New York City. Over thirty life-sized, realistic dinosaurs (including the ninety foot long Argentinosaurus, so big it's visible from the Empire State Building) come to life . . . (Field Station Dinosaurs 2016).

The presence of the ancient past is not cast as an unsettling intrusion which threatens the modern era, nor does the park evoke some straightforward sense of continuity. Rather, the model dinosaurs suggest a sense of participation within a far wider environment. This is part of the mission of the Field Station, which liaises with officials from the local natural history museum to ensure accuracy in their recreated animals, scenes and information panels. This marks a distinction from museum sites where models and replicas of the prehistoric environment are also used. Within the institution, the association with the latest scientific knowledge adds authority to the narrative provided within the exhibition or display (after Foucault 1991: 95). The emphasis on expertise is part of a wider institutional representation which reinforces the identity of museums within public life (after

Macdonald 1998). However, whilst dinosaur parks refer to the way in which palae-
ontologists have worked with their designers to mimic the appearance, movement,
behaviour and sounds of the animals, what is apparent is that this discourse of
authority is not utilised as the defining rationale of the sites. Indeed, present within
all the dinosaur parks is a sense of enjoyment and pleasure which essentially characterises the role of these locations. Whilst education and awareness is promoted,
the concept of fun is what defines the activities within these places:

> At the center of it all is the fun, the joy and the wonder of dinosaurs. Our
> expedition takes every family on a shared adventure – full of mystery, surprise
> and a sense of awe (Field Station Dinosaurs 2016).

The insistence on delight and the joyful encounter with this natural history can be
assessed as an indication of how dinosaur parks represent a 'playful heritage'
(Wilson 2014). Unconfined by a narrative logic drawn from scientific endeavour,
these sites can present dinosaurs from alternative periods adjacent to one another,
provide trails that emphasise excitement by focusing on the biggest and most
famous specimens and include prehistoric mammals within the same complex.
This 'playfulness' is not an admission of relativity, the acceptance of juvenilia
or a disavowal of the intellectual process, rather it is a recognition that there is an
active engagement in the construction of the meanings of the past (after Wertsch
2002). There is a great seriousness in the act of play, as it enables alternative views
and ideas to be formed away from the confining rhythms of everyday society (after
Huizinga 1949). It is within these playful actions that modes of dissent are formed.
Such playfulness is also part of the George S. Eccles Dinosaur Park in Ogden,
Utah. This site, which first opened in 1993, features approximately 100 dinosaur
models which were initially built from concrete and plaster but now represent
the latest in technological developments with robotic elements, new materials
and sound systems throughout the park which are designed to mimic the calls of
the extinct creatures. These scenes are all laid out with the setting of the Wasatch
Mountain Range, part of the large Rockies, which dominates the dinosaur park.
Through the dramatic vistas, provided both by the local geology and the model
dinosaurs, the park is offered to local residents and visitors alike as a 'community
gem' providing an 'educational and recreational benefit to families everywhere'
(George S. Eccles Dinosaur Park 2016). Visitors can walk amongst creatures who
are posed ready to attack, feeding or raising families and regard their presence
within that space. The playful arrangement of models enables a sense of excitement
and fun as visitors can view themselves alongside long extinct animals.

   The use of fun and a sense of playful heritage creates a sense of place around
these dinosaur parks and gardens which alters individual perspectives on contemporary society. From the grand scale sites to the small displays, models of these
animals are used to elicit an emotional response of pleasure and excitement. The
role of fun within these sites is significant as it allows for a play on interpretation
and place. This is not accompanied by a separation from the present but forwarded
as a means of reflection upon the size, scale and dominance of dinosaurs as a

species. Making this ancient history present alongside contemporary visitors places those visitors in an environment where a greater sense of the social can be formed. For example, at Dinosaur Park in Cedar Creek, Texas, the array of replicas has been provided to form a space for a wide, social engagement:

> These statues range in size from the 2-foot long Compsognathus to the 55-foot long Spinosaurus. As you walk through a tree-lined nature trail the dinosaurs sit back from the trail, situated among plants, trees and rocks, making it easy to imagine real dinosaurs in a natural environment. The Dinosaur Park is an educational and fun place, where everyone can learn about the majestic animals that ruled our earth for over 150 million years (Dinosaur Park 2016).

The significance of this can also be seen in the choice of models used within the park. From the *Ankylosaurus* to the *Velociraptor*, both from the Late Cretaceous era, the creatures have been selected to reflect the animals that would have been present in what is now Texas millions of years ago. The trails around the site guide the visitor through various scenarios which orchestrate a sense of surprise as individuals suddenly encounter large specimens or displays of ferocious behaviour. As such, the sense of enjoyment is an obvious aspect of the park's work but so too is a sense of a shared environment. With the focus on exhibiting local specimens, a mission of inclusiveness and an arrangement based on fun and pleasure, the displays at Dinosaur Park can serve as a basis of where humanity learns to move beyond the human.

The playful nature of these parks, where the past is not encountered in a rigidly thematic or chronological order but presented as 'lived' can also be seen within the dinosaur parks in Canada. The traditions of displaying model dinosaurs in Canada has a similar trajectory to what occurs in the United States. From grand exhibitions in the 1930s to roadside attractions from the 1950s, Canadian dinosaur parks are a mixture of the local and the spectacular using both stationary models and animatronics to engage visitors. Calgary Zoo opened its Prehistoric Park in the 1930s and has developed its exhibits over the course of the twentieth century. Today, models are arranged alongside animatronic displays all within recreated environmental and geological settings, which attempt to mimic the various eras from which these dinosaurs evolved. The paths through the 0.02 km^2 Prehistoric Park enable visitors to contemplate the scale of these extinct creatures but also reflect upon the existence of life within this area millions of years ago. This concept of a concurrent time alongside the present is developed further by the use of dinosaur models that would have been part of the ecosystem of what is today Alberta:

> Watch for triceratops grazing among the bushes, then look up to see fearsome T-Rex's banana-sized teeth ready to tear into his next meal. Don't run! You haven't actually travelled back in time – it just feels like it. Discover life-sized dinosaur models around every turn at the Calgary Zoo's Prehistoric Park and see for yourself what Alberta might have looked like when dinosaurs reigned supreme (Calgary Zoo 2016).

The connection with the local and the immediate does not just accord the displays a degree of accessibility. It further emphasises the presence of other lives and another existence beyond the contemporary. It reasserts the presence of the élan vital of these animals as a means of expanding ideas about place and identity in the modern world. Rather than providing a linear narrative from past to present, the dinosaur exhibits succeed in placing chronological periods in tandem. In effect, it casts the contemporary visitor as a social witness, observing the existence of another beyond the immediacy of their own era. Such complex moments of reflection are not limited to the larger parks and gardens displaying model dinosaurs in Canada. Smaller sites also perform this function, providing regional context to this natural history whilst also asserting the significance of wider definitions of the social. Two examples of the smaller dinosaur parks which develop local associations with their displays are The Prehistoric World, Morrisburg, Ontario, and Jurassic Forest, Gibbons, Alberta. Both of these sites provide trails through recreated scenes where model, 'life-size' dinosaurs entertain visitors. At Prehistoric World, a route of approximately 1km in length leads audiences around the site where over 50 dinosaurs from various eras are displayed. Visitors are encouraged to bring packed lunches to eat alongside the animal models as the experience of proximity to the scale and difference of the dinosaurs is part of the appeal of the site. The experience of 'recreated life' in the contemporary locale where once a different world was inhabited ensures the creation of a fundamentally different sense of place which enables visitors to move beyond the human.

At Jurassic Forest, with the use of animatronic models, 'realistic sounds' and a walkway which guides visitors through the enclosures with warnings not to go near the animals, the venue is centred around the visitor experience. As such, it could be argued that the displays render the complexity of natural history down to nothing more than a parade of simulacra – a shallow, facile representation of the actualities of palaeontology and evolutionary history (after Wallace 1996). Dinosaur parks could in this manner resemble the 'hyperreal' (Eco 1986). Formed from a desire to construct the past into a perfect setting of 'realistic' terrains and seemingly 'ferocious' but ultimately harmless animals, such arenas could appear to resemble the 'desert of the real' (after Baudrillard 1994). Reflecting a late capitalist desire to consume, offering a spectacle of the past as a commodity rather than engaging society with reality, dinosaur parks can be construed as symptomatic of the wider malaise within modernity (Lyotard 1984). Indeed, critics may examine dinosaur parks for their apparent pastiche of evolutionary periods and regard the locales as essentially 'non-places' divorced from context and meaning (Augé 1995). However, such assessments would obscure the way in which these sites provide a setting for social performances by concentrating on political alienation rather than public engagement. In considering the latter, it is the performances and perceptions framed by the dinosaur parks that demonstrate a departure from the capitalist, consumer space, as those who critique the processes of late capitalism would define it. It is within the coexistence of past and present exposed by these sites that an alternative vision of nature and culture can emerge (after Žižek 2014). By presenting the occurrence of time, space and place as contingent then the fundamental division between humans and non-humans within the modern age can be assessed (Latour 1993). To bear witness in these spaces is to encounter another understanding of the social (after Latour 2005).

This capacity of the dinosaur park to reflect upon contemporary society is particularly apparent within sites across Europe. Whilst these locales may possess alternative historical trajectories to their counterparts in the United States and Canada, with models and dioramas featuring in public exhibitions from the eighteenth century, the development of commercial parks has emerged particularly since the 1970s. For example, the Dinosaurier Park is an open air museum in Münchehagen, Lower Saxony, Germany, which has developed from the 1980s onwards. With over 200 model and animatronic dinosaurs on display in an area just under 1km^2, it is one of the largest dinosaur parks in Europe. The site is built around a preserved track of hundreds of dinosaur footprints which date to the Cretaceous Era approximately 140 million years ago and which were discovered in the 1980s. Whilst providing access to traces of the past, the park uses a trail over 2km in length to guide visitors through both a geological and palaeontological narrative display. Within these paths is a series of displays where models are used to engage audiences with the sheer difference of the past:

> In the open-air area, visitors can encounter not only all the famous dinosaurs such as Stegosaurus, Triceratops, Brachiosaurus and, of course, the Tyrannosaurus rex but also the most interesting extinct animals of evolutionary history (Dinosaurier Park 2016a).

Whilst the existence of the dinosaur tracks might establish the reality of the creatures, it is the 'hyperreal' reconstructions which offer a means of reflecting upon the cultural formation of reality. The familiar scenes of the dinosaur park, the settings of feeding and family, are present here as are the simulated excavations where children can unearth dinosaur bones and extract model dinosaur fossils from blocks of clay. As such, Dinosaurier Park, whilst focusing on a scientific mission to communicate the history of evolution, uses an approach based on 'exciting moments of experience with practical work' to reveal the presence of this ancient past as a lived experience in the present (Dinosaurier Park 2016b). Visitors are encouraged to take photographs with the animatronic displays and to enjoy the physical experience of walking through the site. Dinosaurier Park emphasises the way in which time is experienced within these spaces as a means of recognising the world beyond the human with its display of prehistoric mammals. Mammoths and a range of other megafauna are exhibited as 'lifelike' models within the park as further evidence of the processes of evolution and extinction. It is within this extension that the way in which humans engage with a far wider world of non-humans can be considered as animals of different modes of adaptation and vastly different chronological periods are placed together. As visitors move around the site, although they may be guided through sections which are structured on notions derived from the establish-ment of climatological or geological phases, it is the collection of specimens in one place that affirms this as a distinct space where alternative lives are witnessed and regarded.

Across Europe, dinosaur parks fulfil the same function. Whilst some are located adjacent to significant fossil finds as a means to enhance the site as a tourist and educational venue and others began as commercial or civic ventures, the significance of these places as a means of reflection is maintained. Saurier Park, near Bautzen in Eastern Saxony, is Germany's other large site for dinosaur models, with over 200 examples arranged along a series of sectors based on chronological period (Saurier Park 2016). The site presents itself as a place of 'excitement, fun and adventure' for children and adults alike so whilst the formal division of the dinosaurs into eras is presented it is the mode of encountering these models which is regarded as significant:

> The giants of Jurassic time like Brachiosaurus, Diplodocus or Barosaurus will impress you by their enormous size. Large predators like Allosaurus, Gorgosaurus and the famous Tyrannosaurus Rex let you feel their former danger. But also many smaller dinosaur species with interesting forms can be discovered and admired (Saurier Park 2016b).

The park thus becomes a place where time periods are placed as coexistent as, whilst model dinosaurs share the same space as humans, the scale of the extinct animals regarded next to the contemporary environment thereby forms a sense of coexistence. Whilst this epiphany might occur next to plastic and fibreglass models of creatures that have not lived for millions of years, the effect is still significant. Indeed, the effect might be amplified because of the extinct status of the dinosaurs. As Mitchell (1998) suggested, the figure of the dinosaur is perhaps the iconic symbol of the modern age, revealing both a desire to dominate and control but fraught with the potential for destruction. Therefore, it is in this engagement at the dinosaur park that humanity can consider its place in relationship to the wider world. The presence of a species that once proliferated across the Earth places contemporary human society in a wider society as it moves visitors beyond a recognition of a synchronic status of the world and an acknowledgement that this place has been inhabited and is still inhabited by humans and non-humans alike.

Such perceptions are affirmed within the dinosaur parks that have arisen in Europe since the 1990s as the public interest in these extinct creatures has increased. In Bałtów, Świętokrzyskie Voivodeship, in central Poland, JuraPark was created in 2004 after the discovery a few years earlier of preserved dinosaur tracks in the nearby mountain range. JuraPark was developed as a tourist initiative for the area as the discovery of the ancient remains drew national and international interest. The site is spread over an area of 0.12km² and is one of the largest tourist sites in Poland. The park is dominated by recreated dinosaur models but a leisure complex, theme parks, restaurants and spas also add to the attractions as do recreated scenes of early human life (JuraPark 2016). The presence of these ventures alongside the recreated settings of extinct life emphasises the aspect of these sites as a palimpsest, formed from the various performances and habits that have defined this space. As a highly popular tourist site within Poland, JuraPark places humanity within a far wider context than its commercial trappings would suggest. Indeed, it is within this context that another concept of society can be considered. The array of tourist sites offering the experience of 'travelling back in time' through the engagement of dinosaur models and animatronics should not be considered a form of capitalist spectacle which obscures the reality of production within society (after Debord 1967). Indeed, it provides a form through which that capitalist spectacle can be undermined and with it the division between nature and culture. It is through the emplacement of people within times and context beyond their own that an alternative society can be considered. This potential, born out of the chronological tensions inherent in the programme of dinosaur parks, can be witnessed particularly within recent sites. In Zator, Małopolska, southern Poland, DinoZatorland opened in 2009 alongside various other leisure attractions as the first site to possess animatronic specimens for visitors to examine (DinoZatorland 2016). However, it is the fusion of past and present, ancient and modern, that marks this space as distinctive:

> The biggest attraction is our nature trail in the old forest, which is rich not only in 'animated' prehistoric reptiles, but also in beautiful centuries-old trees. It is here, where you can admire the world's largest mobile T-Rex! This walk will allow you to move back through time for hundreds of millions of years. You will also be able to listen to the terrifying sounds of dinosaurs. There are also statues of the first people that inhabited our planet (DinoZatorland 2016).

It is in this unique place that well-established notions of place and identity within contemporary society can be challenged. These locales serve as points of formal education for society but their purpose as providing a platform for acts of social witnessing makes them highly significant for wider purposes. For example, Parco della Preistoria in Rivolta d'Adda, Cremona, Italy, which was founded in the late 1970s as a site for natural history instruction, today performs a number of functions from educational resource to tourist site and place of recreation. Over two dozen life-size statues, formed from a range of materials to ensure a 'realistic' appearance, are placed around the 1km^2 site with dinosaurs, ancient mammals and early hominids placed within the parkland setting (Parco della Preistoria 2016). Surrounded by the models of the past, the visitor to the site engages with a world connected and different from their own in the company of others. Therefore, the engagement with this natural heritage is not conducted in isolation, it is always conducted with regard to the social. This is common across European dinosaur parks. The Préhisto Dino Parc, in Lacave en Quercy in the department of Lot in southwest France, also provides a means through which an act of social witnessing can take place (Préhisto Dino Parc 2016). This area opened in 2014 and features a display of dinosaurs along a woodland trail, moving from some of the earliest known specimens to the end of the Jurassic Era. The trail then moves onto Ice Age animals and the evolution of modern hominids before finishing within a recreated Neolithic village. From the origins of life on Earth to the basis of modern society, the visitor is required to regard the points on each new development of species on the pathway as one in the same process. Rather than joined by overt connections regarding evolutionary physiology, these sculptural forms are joined by their shared connection of life on Earth emphasising the significance of the élan vital. Thereby, Préhisto Dino Parc is orientated towards social witnessing where the lives of others beyond the self are experienced, engaged with and understood:

> The dinosaurs, those impressive creatures that have always amazed us and populated our planet 250 million years ago, could reach nearly 15 meters in height and weigh more than 20 tons. During your visit to Préhisto Dino Parc, you can observe as closely as possible these fierce predators. . . . Préhisto Dino proposes to you to discover the lifestyles of these giants, in the very heart of the Lot department . . . territory with multiple traces and discoveries of the dinosaurs and the prehistory of humanity. Dinosaurs will soon have no secrets for you . . . (Préhisto Dino Parc 2016).

In France, dinosaur parks range in style and approach from the Musée-Parc des Dinosaures in Mèze, Hérault department, southern France, which focuses on research and education on a site where fossilised eggs and dinosaur bones were discovered in 1996, providing the impetus for the park (Figure 5.2). Within the Musée-parc des dinosaures, full-scale replicas of animals from across the Mesozoic Era are exhibited within the pine woodland and scrubland of the 0.06km² location. Alongside the trails are display boards featuring fossils and data from the local area which are placed in tandem as a narrative of past and present. The venue also features a display of extinct mammals and reconstructions of the ancestors of early humans which are focused towards engaging society with an era and a perspective beyond their immediate environment:

> In this remarkable site, the public meeting our distant ancestors through recon-structions of scenes of prehistoric life and presentation of exceptional collec-tions: tools, lithic industry, art objects, religious objects, everyday objects . . . (Musée-Parc des Dinosaures 2016).

This 'public meeting' enables visitors to assess the experience of the past as an ongoing concern not condemned to the mists of time. The displays are not used as excerpts of a distant history viewed from the security of our current era but as part of our environment. Divisions of culture, nature and time are circumvented in this process as moving through the park is an exercise in forming a broader concept of the social. This mode of accessibility and engagement is present throughout the park as the display of prehistory, evolution and reconstructed dinosaurs are provided with a sense of the élan vital. The museum's models are not envisaged as inert artefacts that represent the past but points of connection in the present, as the reconstructions are recognised as something that 'moves, breathes, lives' (Musée-Parc des Dinosaures 2016).

Figure 5.2 Musée-Parc des Dinosaures, Mèze, France

The objectives of the operation at Musée-Parc des Dinosaures, which was built around the discovery of fossilised remains, can be considered alongside venues in France which have emerged from private interests. For example, Cardo Land in Chamoux, within the Yonne department of central France, was created by the artist and filmmaker Cardo (Raoul Cardo Saban Torres Irigaray) (1924–2009) in 1984.

On this site, dinosaur models have been created from metal and cement with the snarls and growls of the creatures audible to visitors as they walk through the 0.08km² country park. These recreations are accompanied by models of prehistoric mammals such as mammoths and sabre-toothed tigers as well as sculptures of early hominids (Cardo Land 2016). The Parc Préhistorique de Bretagne in Malansac, which opened in 1988, is organised upon a similar structure. Visitors are guided through thirty life-like scenes of dinosaurs within a 'natural' setting before being provided with scenes from the development of human and animal life on Earth. From the Palaeolithic onwards, models and reconstructed habitats are used to engage visitors with a prehistory as 'they have never seen it' (Parc Préhistorique 2016). Therefore, no matter the scale or impetus for development of these dinosaur parks, whether scientific discovery, tourist initiative or private ambition, these locales provide a social engagement with the past – an act of witnessing that necessitates a consideration of the lives of others. In this sensuous engagement with natural heritage, another mode of being can potentially emerge:

> By reason of this with-like Being-in-the-world, the world is always the one that I share with Others (Heidegger 1962: 155).

The capacity of sites to evoke such sentiments is reflected in the layout and design of these spaces. Forming points of connection within guided trails through sculptural groups and recreated scenes ensures a vision of these sites as very different places, where well-held notions of time and space are transformed as an alternative vision of society, can be considered. In Britain, this process is present within the array of dinosaur parks across the country which also reflect the private, public and commercial character identified elsewhere. From additions to safari parks, garden centres, nature reserves or tourist sites, dinosaur and other prehistoric models provide a point of connection with an alternative environment and a means to be a social witness, or a Being-in-the-world that I share with others (after Heidegger 1962). Within these venues, visitors are frequently reminded that they are sharing a world with another creature which disrupts their place as the dominant species on the planet. This is demonstrated in the juxtaposition of one of Britain's largest animatronic dinosaur parks adjacent to a wildlife enclosure in Bewdley, Worcestershire. The Land of the Living Dinosaurs was built as an extension to the West Midland Safari Park and opened in 2015 (Figure 5.3). The site possesses just under 40 animatronic models which have been formed from rubber and metal and whose moving necks and jaws are designed to mimic 'life-like' action, set against a landscape background of rock formations and planting schemes designed to evoke the various ages of the Mesozoic Era represented within the venue. From dominant displays of a *Tyrannosaurus rex* which watches visitors from a rocky outcrop to the dramatic reconstruction of a 30m *Argentinosaurus*, the site provides the exciting and engaging atmosphere present within the dinosaur parks in the rest of Europe. The sense of Being-with other times and other species as a process of Being-in the present is used as part of this natural heritage throughout The Land of the Living Dinosaurs (2016). This is most clearly represented in the playful

Figure 5.3 The Land of the Living Dinosaurs. West Midlands Safari Park.

use of warning signs dotted around the venue which caution visitors that the attractions are more than just models:

> Keep to the Paths: If not, the dinosaurs may eat you and that might make them sick (The Land of the Living Dinosaurs 2016).

It is in such confrontations with humanity's place within the world that dinosaur parks reveal the arbitrary boundaries between nature and culture within contemporary society. As visitors are placed within a time and place where humanity's existence and position on Earth is made into a contingent issue rather than a certainty, the space for an alternative vision of the modern world can be formed. The presence of live animals within the safari park next door emphasises the role that the fabricated models have within such an environment. The living animals constitute an object of observation, fixed by the gaze of the human visitor as part of the natural world apart from humanity. The dinosaur models and their recreated scenes constitute a means of placing humans back into the company of nonhumans, not as an exercise in voyeurism but as a process of making an alternative experience of the environment. Similar arrangements of living animals within parks or venues providing dinosaur models can be observed at another well-known site in Britain, Dinosaur Adventure in Lenwade, Norfolk, Eastern England. Within this 0.3km² site, models of extinct creatures from the Jurassic Era are part of a series of attractions that include farm animals and a deer safari (Dinosaur Adventure 2016). The display of extinct and living creatures alongside one another ensures the act of social witnessing within these sites is performed and enhanced. It is within these places that the modernist distinction of nature and culture is both affirmed and then dislocated.

It is through the dinosaur parks that an alternative perception of the environment can be formed where humans can consider their place within the world and their relationship with other non-humans. Such places form locales of social witnessing as individuals within these sites are made to reflect and reconsider the issues of temporality, space and identity which forms the basis of modern western society (Giddens 1990). If a different means of thinking about the human and the non-human is to be formed, then it must be created within spaces that were not constituted by those established forms of time and space. What is required is a disruption to those values and a recognition that any attempt to confine society to a set of parameters will be framed by those established practices. As such, the relationship between nature and culture within western society can be reconsidered on the basis of alternative, fluid and renegotiated assessments of time and space (after Laclau 1990: 92). This is the format that dinosaur parks provide, a trail through time and space which, whilst perhaps ordered by chronological era, relies upon a process of excitement, engagement and playfulness rather than order and categorisation. It is not the subject matter of dinosaurs that enables this perspective and mode of witnessing to occur. Similar theme parks or leisure sites which might use geology, prehistoric mammals or human evolution can achieve the same effect of disrupting how past, present and future are imagined. What is significant is that these spaces form a very different type of place where visitors can observe not the distancing of humans and non-humans but their co-existence. Humanity might well be relegated to a participant within a far wider notion of society in this scheme but the human becomes more secure in the process. It is within the company of others that a clearer assessment of humanity's role as part of the environment can become a matter of concern.

The moral gaze: natural history on film and television

Over the past 70 years, the screen has become the most significant frame through which a sense of natural heritage is communicated to the wider public in the contemporary western world. The success of the film 'Le Monde du silence' (1956), directed by the underwater explorer Jacques Cousteau (1910–1997) and the auteur Louis Malle (1932–1995), has been cited as the origin of natural history television with its dramatic capture of wildlife. From the early 1950s onwards, television companies within Europe and the United States have drawn upon this sense of drama and excitement as a means of popularising natural history and educating the populace (see Davies 2000a). However, the representation of natural history has a far longer history on screen, with some of the earliest motion pictures using the subject as a powerful form of engagement to capture the imagination of the audience.

The grand scale of natural history appeared to be well suited to the spectacular new medium of the cinema. From the early twentieth century onwards, adventure and romance films, either original productions or book adaptations, attracted viewers through the depiction of the sublime quality of natural heritage. Through either extinct animals, geological processes, climatological events or astronomical anomalies, the fearsome quality of nature was used as a plot device within these films. Faced with the dangers of disruption and change from the natural world, this heritage was frequently cast as an object to be overcome and to be survived.

Therefore, it is possible to note two distinct traditions in the representation of natural history on film and television: as a means of popularising science and as a means of popularising nature. The former is concerned with providing detail, information and narrative to the processes of the natural world whilst the latter places character, form and meaning onto the environment. In recent years, scholars have emphasised the divisions between these two approaches as a richer and more nuanced understanding of natural history is promoted to combat the dangers of environmental degradation and climate change. However, the two traditions do not represent alternative aims and objectives; rather they utilise different methods of achieving these goals. Both of these approaches situate the viewer in a particular relationship with natural heritage; they require the individual to serve as a moral witness to the dangers, effects and ultimately the fragility of the environment. The representation of natural heritage on film and television provides a means for shaping witnesses who are asked to hold ideas, values and identities towards the object of their viewing (after Wheatley 2004). This media representation provides a means of constructing moral witnesses as an active engagement with these cultural forms (after Hansen 1991; Perkinson 1991).

The role of natural history programming on television has developed over the course of the twentieth century, with new techniques, narratives and perspectives being provided by technological advances or scientific developments (see Cottle 2004). This has ensured the development of particular aural and visual aesthetics within natural history television programmes which have come to dominate the agenda of production companies within Europe and the United States (after Davies 2000b). Indeed, whilst the content and narration of natural history programmes might differ within nations, increasingly natural history on television is becoming internationalised with large budget productions shared across a number of broadcasters (Richards 2013). In Britain, natural history on television has been dominated by the productions of the state broadcaster, the British Broadcasting Corporation (BBC), as well as private television companies (Boon and Gouyon 2015). Indeed, one of the most nationally and internationally successful natural history documentaries was made by the company Anglia/ITV, which from 1961 to 2001 produced the series 'Survival'. However, since the 1950s, the BBC has operated a Natural History Unit based in Bristol, southwest England, which has provided innovative and highly popular programming covering all aspects of life on Earth in the past and the present (Bassett, Griffiths and Smith 2002).

The success of the Natural History Unit over the past few decades has led to criticism from the academic sector regarding the accuracy of reports and the

inevitable exclusion of some aspects of the scientific process in the editing of programmes for the public (Jefferies 2010). Certainly the public mission of the Natural
History Unit, which follows the wider role of the BBC to 'inform, educate and
entertain', is focused on the development of visually stunning and engaging narratives to raise interest and awareness of natural history (BBC 2011). This function
can be observed within the subject and tenor of the natural history programmes
that have been produced since the 1970s (Scott 2010). These programmes have
addressed varied geographical areas and multiple subjects and can be seen as
points of reflection on contemporary issues (see Berry 1988). From concerns
regarding conservation and environmentalism to climate change, the output of the
Natural History Unit has been used to move audiences towards reflection on
the dangers faced by animals and the wider planet. Surveying a sample of the
output of the Unit also demonstrates the extramural uses of natural history programming. Examinations of the natural history of China, India, Japan, Brazil and
the United States appear to be broadcast to coincide with major geopolitical
incidences or alterations. Regardless of period, agenda or region, what is common
throughout this programming is the manner in which the audience is required to act
as moral witnesses to the issues before them. Through the frame provided by the
television screens, viewers are required to express concern with the subject matter
presented as scenes of natural history, its grandeur, its tragedy and its importance
are relayed:

- Life on Earth (1979) – In this first major BBC nature documentary, which
  became an internationally renowned, landmark television series, the evolution
  of life on Earth is presented as a series of struggles and eventual successes. In
  episodes entitled 'Conquest of the Waters', 'Invasion of the Land' or 'Victors
  of the Dry Land', the development of life is portrayed as a series of advances
  culminating in the arrival of humans.
- The Living Planet (1984) – This account of the geological formation of the
  planet, alongside the development of the Earth's climate and flora, emphasised
  the importance of protecting the ecosystem. The final episode, entitled 'New
  Worlds', highlighted the damage wrought by humanity as it has fundamentally
  altered the planet and the need to establish a sense of equilibrium.
- The Trials of Life (1990) – This examination of the natural history of animal
  behaviour focused on the individual and social actions of animals as they
  grow, feed and reproduce. This behaviour is presented within an anthropomorphic setting, with episodes entitled 'Home Making' and 'Living Together'
  providing a stunning image of animal life as a mirror of human existence.

The major documentaries such as these paved the way for an array of programming
which has focused on providing viewers with visually powerful depictions of
natural history (see Elliot 2010). Indeed, the BBC programmes also provided
a stimulus to other natural history programmes across the world as the Natural
History Unit's documentaries were licensed to other countries. In the United States,
PBS (Public Broadcasting Service) began screening the nature documentary 'Wild

America' in 1982, which focused on the wildlife of the North American continent. Whilst this programme was broadcast until the early 1990s, the documentary series 'Nature' featuring wildlife and ecological reports is still a part of the schedule on PBS after its first series in the early 1980s (see Bousé 2000: 81). Whether examining contemporary contexts or prehistoric eras, the function of these programmes is consistent. Natural heritage is formed as a shared object of reflection, a mediated site where individuals and communities alike can obtain a sense of their own values and identity. It is within these documentaries that an act of moral witnessing is undertaken:

Life in the Freezer (1993)
Land of the Tiger (1997)
The Life of Birds (1998)
Living Britain (1999)
Walking with Dinosaurs (1999)
State of the Planet (2000)
Blue Planet (2001)
Walking with Beasts (2001)
The Life of Mammals (2002)
Galápagos (2006)
Wild China (2008)
Ice Age Giants (2013)
Life in the Air (2016)

This is a highly important function of these programmes. To cultivate a sense of moral witnessing enables wider society to regard the environment as an aspect of their lives. Indeed, such acts of witnessing create a space for this concern within contemporary life. Watching these programmes allows individuals to form, practice and cultivate moral responses to the wider environment (after Dant 2012). Media representations of natural history do not provide a distanced, alienated understanding of the habitats of other species; television is not a medium that further divorces humanity from nature (after Hartley 1999). In fact, television programmes serve to vividly connect society with the lives of non-humans by demanding a recognition of their existence (after Frosh 2006). The nature documentaries created by the Natural History Unit of the BBC establish the moral imperative that flora, fauna, palaeontology, geology and climatology be recognised. However, it is in this action that the limitations of moral witnessing can be realised, as the form does not necessarily convey any other mode of understanding humans and non-humans beyond the modernist divisions. Indeed, it can be noted that, whilst the viewer observes the processes of natural history, this is always accomplished with the perspective that the object of this concern is 'the other'. In effect, whilst the medium of television might provide closer engagement and vivid illustrations of the complexity of environment, it inevitably frames this relationship in a particular way (after McLuhan 1964). It does not divorce humans from non-humans but it does serve to cultivate the distinction between nature and culture. The television allows society to witness the epic environmental events and the miniscule processes that underlie life on Earth but this is conducted as a world apart. It reaffirms the distinctive spheres in which the human and the non-human are assumed to exist whilst providing a powerful means of engagement.

This normalises the modern relationship between nature and culture and constitutes an affirmation of the perspective of the moral witness as viewers are obliged to recognise a role and responsibility towards the object of their vision. This affirms the notion of 'media witnessing' as described by Frosh (2011), where television demands an acknowledgement of the other that we witness but always from a detached level. In essence, this can be classified as a process of 'civil inattention'; acknowledgement is extended to the other but its purpose is to maintain the processes and structures of society (after Goffman 1972). 'Civil inattention' ensures a degree of stability as individuals and communities preserve the same patterns of behaviour and action. Nevertheless, in this assessment, civil inattention allows the acknowledgement of environmental or conservation issues through the representation of natural history, but this is focused on the exercising rather than the challenging of the moral gaze. Ultimately, the same structures within society that threaten or imperil the contemporary world remain in place. The application of this assessment to natural heritage and television can be assessed as a process of moral witnessing or 'media witnessing' which is characterised by the evocation of concern, an affirmation of social boundaries and obligations but not necessarily an alteration in action (Frosh 2011):

> As a moral force, then, media witnessing – like civil inattention – is a routine and institutionalised social procedure for moralizing strangers . . . (Frosh 2011: 68).

Media witnessing or moral witnessing can be classed as the same process of engagement with the subject represented through the television, which normalises social and civil inattention whilst simultaneously ensuring access and engagement (after Frosh and Pinchevski 2008). In the context of natural history programmes, the mode of witnessing present within these formats also becomes a means of establishing common concerns across different nation states. As successful natural history programmes are frequently supported by international collaboration and highlight the threat posed to the environment, the television forms a 'risk-cosmopolitanism' (Beck 2006). Viewers of television across the world are brought together as the dangers faced by contemporary society are detailed before them. However, this is a community of civil inattention where what emerges is a reiteration of the importance of caring and compassion but which is always defined by the existing frameworks within society. Whilst television ensures a level of engagement, it is an engagement formed through the established separation of nature and culture. It forms a strong means of identification within human society but its involvement within non-human society is limited because of this connection.

Natural history programmes within Europe follow this same pattern of division and civil inattention through the use of the moral witness perspective. Viewers are provided with a means to establish a testimony that evidences their concern and engagement as images of natural history exercise and re-establish a shared moral vision. In France, this is particularly noticeable in the way in which historical television documentaries have provided a means to establish both national and cultural ideals (after Veyrat-Masson 2000). The representation of natural history on television within France has been a prominent part of educational and scientific programming over the past sixty years (Garçon 2005). In recent years, natural history programming within France has developed with extensive documentaries and film providing engagement with this heritage. This builds upon the established place of documentaries within French television culture provided by such series as 'Histoires naturelles' (1981–2009), which has been broadcast since the early 1980s. The series finished production in 2009 but is still to be found on television on the former state-owned channel TF1. The series was begun with the express desire to connect humanity and nature in an era when a number of commentators within France had highlighted the disconnection between the modern world and the patterns of tradition (see Nora 1984). 'Histoires naturelles' (1981–2009) featured both local and national environmental history from France but also covered the processes of natural history across Africa, Asia and South America as well. Over the course of its broadcast history, the series increasingly moved towards the use of natural history as a means to examine the effects of pollution and environmental damage inflicted by humanity. As such, the position of the moral witness was affirmed as the long-running programme asserted the significance of looking upon nature as a means of demonstrating a shared vision of the world. The witness is required to acknowledge an obligation of holding a moral response to what is depicted, but such actions are not extended beyond the connective point between the viewer and the screen.

This civil inattention in the midst of programmes of natural history which otherwise offer an evocative and vivid description of this heritage is noticeable within other more recent programmes in France. Since 2009 the documentary series 'Grandeurs nature' (2009) has been broadcast on the state broadcaster France 2. This has provided viewers in France with a range of studies on natural history from across the world. Significantly, these programmes have been provided as a means by which individuals and communities can connect with the environment but also take a moral stance on issues of conservation and protection (see France 2 2016).

The exercise of the moral witness is evident here as the nature documentary provides a means of connecting with the plight of species threatened with extinction and the problems that beset our world. This forms a clear example of the 'risk-cosmopolitanism' defined by Beck (2006) which draws societies together through a shared response to perceived threats. The distribution of documentaries and films through the international channels such as National Geographic and Discovery further evidences this perspective. Potentially, this collective response might provide a means of developing new ideas about the environment but the effect of the moral witness ensures that these are values held rather than practices changed (after Hirschauer 2005). Other natural history documentaries in France which focus on the ancient past similarly affect these responses where the morality of the viewer is extended but social norms are reinforced. For example, in 2003 the second state-owned broadcasting channel France 3 launched a highly regarded, major documentary series by the filmmaker and director Jacques Malaterre, 'L'odyssée de l'espèce' (The Odyssey of the Species) (2003). This production fused a traditional documentary style with dramatic reconstructions in its examination of hominid development and evolution over millions of years. The programme explores the unique status of humanity, detailing the emergence of the genus out of Africa and the spread of species of hominids throughout Asia and Europe, before concluding with the dominance of modern humans across the world. Whilst the object of the moral witnessing in this programme is both physically and chronologically distant, the effect is no less diminished (Gauthier 2011). Such programmes which detail a narrative of human evolution enable a moral perspective which affirms an inclusive narrative in the present as individuals and communities today observe the struggles, triumphs and commonality of humans in the past.

Similar documentaries have been developed in France over the past decade, on a variety of themes of natural history, which fuse the traditional model of the documentary with reconstructions which are either staged or computer-generated. For example, the follow-up to 'L'odyssée de l'espèce' was 'Le Sacre de l'homme' (The Rise of Man) (2007). This documentary ,which first aired in 2007 on France 2, narrated the history of humanity's relationship with the environment. 'Homo Sapiens' (2005) used a narrative element with characters, plotlines and reconstructions to present a documentary drama on the lives of the first modern humans, which was shown on France 3. Programmes featuring dinosaurs or extinct animals are frequently either collaborations between French and international television companies or they are syndications of programmes dubbed into French from Britain and the United States. The provision of natural science documentaries from organisations such as National Geographic and the BBC demonstrates the 'risk-internationalism' present within this use of heritage. For example, the 2012 documentary 'La vraie vie des dinosaures' (The True Life of Dinosaurs) (2012), which was broadcast in France with the launch of the digital television channel 6ter, was produced by the Spanish television channel Explora. This particular documentary focused on adaptation and survival of the species (see 6ter 2016).

The moral witnessing within this type of natural history programme serves to reinforce a shared perception of equilibrium with nature as well as the potential doom that might befall any species on Earth. The representation of the risks of life within these documentaries contributes to a shared awareness regarding the threats to the continued existence of humanity on the planet. Such collective acts of moral witnessing ensure that television becomes a very powerful medium for forming values and opinions which inevitably replicates moral judgement rather than establishing alternative practices. Natural history documentaries across Europe perform this function within contemporary society. For example, the Spanish state broadcaster, Corporación de Radio y Televisión Española (RTVE), and regional broadcasters have shown natural science and environmental programmes from the 1990s which affirm the role of moral witnesses (after Montaño 1999a). For example, the Andalusian television company Canal Sur produced the series 'Tierra y Mar' (Land and Sea), 'Pobladores del Planeta' (People of the Planet) and 'Espacio protegido' (Protected Spaces) since the 1990s, which all focus on the relationships between people and places in the region (Montaño 1999b). This mode of environmental programming has been highly significant within Spanish broadcasting, leading to series which provide a vivid account of how humans and nature have interacted in the past and the present.

This can be seen from 2004 to the present in the documentary series 'Espacios Naturales' (2004), which was broadcast on La 2, the Spanish state-operated channel. The wide-ranging programmes which aired under this title provided viewers with a means to observe the variety of natural history in the country. This shared, collective vision is significant, as it does foster a sense of engagement and identification with the environment. It also forms a place where the moral witness can observe and testify. Through such programmes, the role and responsibilities of civil society are affirmed (see RTVE 2016a). This was continued in 2016 when La 2 also aired the series 'Red Natura 2000, la vida en los espacios protegidos de España' (Nature 2000 Network, life in the protected areas of Spain) (2016), which was associated with the European Union initiative Natura 2000. This series was commissioned to highlight the lives of communities across the specific regions which had been placed under conservation orders. The series focused on the interrelation between people and the environment as issues of sustainable development were made prominent (RTVE 2016b). It is within the expansion of environmental programming in Spain from the 1990s that the position of the moral witness is confirmed. This can succeed in raising the awareness of environmental issues and the importance of conservation. However, it is also based on the premise of civil inattention through a mediated form. By ensuring the use of the viewers' moral vision, a greater consensus on the value of environmental work is obtained but not translated into moral action, as established boundaries are affirmed.

These divisions between nature and culture are present within natural history programming across Europe. In Germany, the popularity and spread of television ownership in the West was accompanied by a developing interest in natural history programming which has ensured strong audience figures for documentaries (after Hupke 2015: 29). One of these established natural history programmes is 'Expeditionen ins Tierreich' (Adventures to the Animal Kingdom). The series was first broadcast in 1965 by the regional television channel NDR and still has episodes produced under the programme title (NDR 2016). The series, whether through new instalments or repeats of the older programmes, provides viewers with demonstrations of natural history from across the world that extend the moral vision but also distinguish the separate roles of nature and culture. With each episode, the sights of nature are cast as objects of concern and fascination for the audience. A similar process can be observed with the current natural history programmes produced by NDR. 'Wilde Heimat' (NDR 2016) (Wild Home) was broadcast after 2005 and connects viewers with the environment and natural heritage across Germany and Austria. Similarly, 'NaturNar', which has been part of NDR's programmes since 1998, highlights the relationships between people and the environment and the importance of conservation. This emphasises the paradox of the moral witness, as in the moment when nature becomes accessible, shared and communicable through representation television it remains distinguished from human communities. This can be discerned within the contemporary social and natural documentary programme 'Terra X' which has been shown on the state broadcaster ZDF since the early 1980s. The range of outputs under this particular programme is significant but it is in the overarching theme of exploration that the natural and cultural divide is made. The human interest is represented as inquisitive, an active assessment of the world, whilst nature, though dynamic and engaging, remains the object of concern (ZDF 2016a).

This process of distinction has been made within the wider natural history output across Franco-German television over the past few years which includes series produced or co-produced by German broadcasters including ZDF, NDR, WDR or the European network ARTE. Shared across the continent, these programmes provide a common means of engagement but significantly also affirm the role of nature and the non-human as distinct from human communities. Whether it is connecting viewers with the ancient past, the chronologically remote, the importance of conservation and the excitement of discovery, the non-human is represented as a shared system of value judgement:

- Wildes Deutschland (Wild Germany) (ARTE 2011)
- Die Odyssee der einsamen Wölfe (The Odyssey of the Lone Wolves) (ARTE 2016a)
- Frauen und Ozeane (Women and Oceans) (ARTE 2016b)
- Geheimnisse der Tiefsee (Secrets of the Deep Sea) (ZDF 2016b)
- The Desert Sea (ZDF 2016c)
- The Amazon of the East (ZDF 2016d)
- Planet Deutschland – 300 Millionen Jahre (Planet Germany – 300 Million Years) (ARTE 2015)

As we gather as witnesses to acknowledge and testify as to the moral importance of understanding and protecting the natural world, we perpetuate the conditions of society which have led to the damage and endangerment that we recognise as potentially disastrous. Similarly, we can also bear witness to the damage wrought by environmental incidents such as earthquakes, landslides, extreme weather and volcanic activity in the variety of 'disaster' television documentaries that feature aspects of natural history and which have become prominent in Britain, Canada and the United States over the last twenty years (see Campbell 2013). These programmes provide another point of moral witnessing where the effect of these events and the wider natural history is examined as a process of human tragedy alongside the 'spectacle' of the sublime environment (see Lester and Cottle 2009). As such, audiences are able to draw moral lessons from such depictions by witnessing the devastating impact of such forces of nature upon humanity. The division between the human and the non-human is reinforced in this manner as the viewer becomes a witness to the threats and risks posed to society. Indeed, as individuals and communities are called upon to testify as to the loss and trauma of environmental disaster as represented by documentaries or by news reports, the recognition of natural history within these accounts serves to reinforce social norms and duties (after Weik von Mossner 2011).

This perspective is continued across both television and cinematic represent-ations of natural heritage. Films which draw upon the environment or natural history do so to reflect upon moral positions as this heritage is used as a plot device or character within the drama. Indeed, from the earliest period of cinema in the late nineteenth century and early twentieth century, natural history has been used to cast a moral vision for viewers (see Pick and Narraway 2013). The earliest repre-sentations of natural history provided a sense of the power and brutal force of nature as it was inflicted upon humanity. In Britain, 'Rough Sea at Dover' (1895), one of the first natural history films to be publicly available, features the powerful waves crashing against the quay at the southern English port. As cinema developed from the early twentieth century onwards, two distinct branches of natural history representation on film developed: the educational and the entertainment (after Bousé 2000). The former has seen the distribution of informative documentary films in Britain and the United States; for example, features such as 'In Birdland' (1907), 'Urban Science: The Birth of a Flower' (1910), 'Simba: The King of the Beasts' (1928) and 'The Private Life of Gannets' (1934) paved the way for the natural history films of Jacques Cousteau from the 1960s onwards which found a global audience. Cousteau's films, which included 'Le monde sans soleil' (1964) (The World Without Sun) and 'Voyage au bout du monde' (1976) (Voyage to the Edge of the World), encouraged natural history programmes on television as the small screen largely became the preserve of educational films and document-aries (see Mitman 1999: 205). In contrast, the cinema became the location for films which provided a popular and entertaining, though not always accurate, vision of the of the natural world. However, despite the demarcated territories, content and altering styles of representation, the perspective of the moral witness is affirmed within both these formats.

From the depictions of early hominids and dinosaurs in early twentieth century cinema in the United States through such films as 'Man's Genesis' (1912), 'His Prehistoric Past' (1914), 'Brute Force' (1914), 'Tarzan of the Apes' (1918) and 'The Lost World' (1925), nature is represented as a challenge to humanity or as an arena where values and ideals are tested. Even in comedies, such as Buster Keaton's 'Three Ages' (1923) or Stan Laurel and Oliver Hardy's 'Flying Elephants' (1928), the environment is placed as a means by which humanity's perspective is tried. For example, within 'King Kong' (1933), the brutality of the natural world is contrasted with human society's own capacity for cruelty. The earliest natural disaster films also extend this moral vision of the cinema with 'San Francisco' (1936), 'The Hurricane' (1937) and 'The Rains Came' (1939) detailing the damage inflicted upon the world by earthquakes and extreme weather. The role of natural history as the inspiration for catastrophe has been a staple part of cinema in the United States in the latter twentieth century. Examples include 'The Devil at 4 O'Clock' (1961), which focused on the devastating human costs of a volcanic eruption, or the attempt to survive the catastrophe unleashed by seismic activity in

Los Angeles in 'Earthquake' (1974). These films provide points of reflection on humanity's relationship with one another, as viewers were required to affirm moral norms in the face of environmental disaster (Keane 2001: 17). Indeed, the films depicting elements of natural history, whether its capacity as a destructive element in human society or natural history as a means of moral reflection, ensure that this cultural representation has been framed by a concept of process throughout the century.

Natural history features in film as a motion-image, a constant process of teleological movement which fixes a sense of the inevitable (after Deleuze 1986). From the earliest films to the modern blockbusters, natural history provides the image of development; whether encompassed by life-threatening lava flows, hunted by fearsome dinosaurs or left devastated by earthquakes, natural history is a means of ordering time, space and experience. As such, the moral lessons it induces reflects the significance of stability and order. As the dinosaurs of 'Jurassic Park' (1993), the volcano in 'Dante's Peak' (1997) or the earthquake in 'San Andreas' (2015) represent a challenge to the established process of normality, the resolution is a return to the previous mode of existence. The moral witness to these accounts regards the presence of natural history as a violent disturbance which is resolved through the maintenance of order. As such, whilst the natural world is quite powerfully evoked within these films, indeed, made disturbingly present through such depictions, it is an image of motion disrupted but which returns to a stable course of development (after Deleuze 1986: 6). Intriguingly, there exists no challenge to the representation of time and space within films depicting natural history. It would appear that the processes represented by nature necessitate the reiteration of structure. The natural 'forces' that are represented within these films are monist, devoid of the wider complications and contexts that are formed through a notion of a human and non-human society. As such, nature and culture are distanced from one another as the former is observed to be a representative image rather than an active mode of expression (after Deleuze 1989: 142).

Therefore, the films featuring natural history are highly moral, they invoke issues of responsibility and justice, but it is the establishment of process, the confirmation of those norms, that ensures that the witness testifies to maintenance of order and the distinction between nature and culture. Rather than reflecting a disparity between high and low culture, the documentary and the motion picture which represent natural history actually stem from the same point of origin and assert the same position. Whether it is a documentary exploring the evolutionary development of animals, the life-cycles of plants, the extinction of species or the geological processes that shaped the planet, the habits and values of human society are reasserted rather than challenged. The moral witness testifies to the ideals that are represented within these programmes rather than assesses the structures of society that have imperilled the environment. Films depicting scenes or subjects of natural history operate in the same manner by emphasising nature as a singular 'force', whether that is earthquakes, volcanoes or dinosaurs, rather than as an aspect of society which is composed of various components and which exists in various points in time. Television and film render natural history into a vivid and engaging spectacle. However, in that point of accessibility, the separation between the human and the non-human is affirmed and the processes of civil inattention are continued.

Natural heritage in language and memory

The form of engagement with natural history across western society is revealed through the intangible heritage of language and memory. Whilst museums, statues, installations and parks constitute the material presence of this past, it is in the way natural heritage constitutes part of a wider set of social and commemorative discursive practices that structures place, politics, power and identity within the modern world (Wilson 2016). It is through language, metaphor, reference, allusion and simile that natural heritage is made present within western society. Across European and North American societies, environmental history is evoked within the political, public and media sectors. Whilst the geological processes and extinct animals may be far removed from everyday experience, contemporary society is still surrounded by dinosaurs, mammoths or Neanderthals, or still experiences the Ice Age or the Jurassic, as species and events are used as references to convey attitudes and ideas in the present. This use of language is more than just for illustrative effect (see Goatly 2007). Indeed, it reveals the relationships that contemporary society bears towards natural history.

It is within this engagement that the manner in which a modern perception of how the world is structured can be understood (Gadamer 1975). It is through the metaphors and references that society conveys its fundamental understanding of the wider environment (Heidegger 1962: 100–101). As we compare the external world through various frames of cultural association, we expose the attitudes we possess to the past, present and future (Lakoff and Johnson 1980). In examining contemporary discourse within Europe and the United States, the effect of these references can be observed to be the formation of political witnesses. Whilst the use of associations and similes to natural history ensures that society lives with this past and that this heritage forms part of wider debates regarding power, place and identity, it is nevertheless equated with a regressive and limiting perspective. Natural history is evoked within language as a measuring device for the modern world, both physically and ideologically. We use natural history to condemn our failings and to emphasise our progress. As such, we become political witnesses with such a discourse as we affirm the development of the present and use it to direct the future (Giddens 1990). When we speak of natural history as a reference, we confirm our place within the modern world (after Latour 1993).

This assertion of the values of modernity and the distinction between the human and the non-human is derived from the legacy of the intangible heritage of language and natural history. The terms and phrases that are employed within contemporary society are derived from the study of the environment during the age of the Enlightenment from the eighteenth century onwards. Names and processes, which emerged from the early scientific discourse of geology, palaeontology, botany and climatology, became part of a popular lexicon within several languages by the twentieth century. The transition from academic to popular usage represents the establishment of a 'collective memory' where the shared vision of natural history from the events of millions of years ago to the contemporary era is formed through such uses of discourse. Intriguingly, the manner in which this heritage is evoked within metaphor, allusion and simile is overwhelmingly negative. Natural history is cast as an intrusion into the modern age which threatens the advances of society. As such, to speak of natural history is to form an act of political witnessing which asserts a separation between the human and the non-human.

This can be observed with the usage of 'dinosaur' within contemporary discourse. The term was coined by the biologist Richard Owen (1841) and was used to classify the fossilised remains of species which were characterised by their size and their disappearance from the Earth. Early interpretations for this extinction focused on the apparent inability to adapt and the cumbersome scale of these creatures. Therefore, the perception of dinosaurs as dim-witted and backward marked the scholarship of the nineteenth century and coloured the use of the term within wider society. By the beginning of the twentieth century, popular usage of the term dinosaur had become a means of derision or critique as it was applied to those who were regarded as outdated, irrelevant and a barrier to progress – in essence, those who were not part of the modern age. Such uses persist within the popular media and they are applied in a variety of circumstances, from gender politics, business and economics to the responsibilities of the state. A cursory glance at British newspapers over the last ten years demonstrates the function of this discourse; politicians regarded as outdated, political viewpoints assumed to be unfashionable and political careers appearing to be at their end are all framed alongside the term 'dinosaur' (Anon 2016a; Merrick 2008; Lyons 2010).

Similar uses exist within the United States and other Anglophone countries. The context and background of these examples is almost irrelevant as the term can be applied to any circumstances where the forces of modernisation are being confronted by the perceived barriers of traditionalism. However, what is evident within these utterances is the existence of 'dinosaurs' within contemporary society millions of years after their extinction for a particular purpose. The employment of these references could indicate the creation of a wider society where human and non-human are participants. The appearance of 'dinosaurs' might demonstrate a greater engagement with natural history than is sometimes assumed. Such potential for an alternative model of society which is facilitated by language is disrupted by the recognition of the negative connotations that are relayed within this usage. The epithet 'dinosaur' exists across a number of languages as a means of rebuking those enthralled to tradition and withholding change. In France, to act or appear outmoded or to seem inefficient or not fit for purpose brings with it the accusation that an individual, group or company is behaving 'comme un dinosaure' (like a dinosaur). Whether it is officials, institutions, corporations or individuals, media representation of inefficiency or hesitation to reform is likened to the animals that once dominated the Earth for millions of years. For example, in a recent report regarding the national education system in France the animals were evoked as the schools were assessed as outdated dinosaurs (see Nunès 2015). Similarly, in a discussion regarding the future of the French national train service SCNF, the state-owned institution was regarded as a 'dinosaur' (Béziat 2016). Coverage of the 2016 Presidential Election in the United States also brought the usage of this term to the fore as a mode of critique, as Donald Trump was equated with the extinct animals for his comments about women (see Bourcier 2016).

The context of the application of this reference is less important than the message it conveys; through this association, natural history is cast as irrelevant for the modern age and a barrier to progress. The emergence of this comparison in French in the early twentieth century has provided a vivid political discourse over the past century and remains an important element of critique. To be labelled as 'un vieux dinosaure' (an old dinosaur) or to be a member of 'les dinosaures de l'académie' (the dinosaurs of the academy) or 'les dinosaures de la politique' (the dinosaurs of politics) is to be ostracised from the wider function and performance of society. Whilst the term can denote a figure of standing within a particular sector, it would also be accompanied by the implication that this position has been overstayed. Those who are regarded as being unsuitable for the present because of their inefficiency, opinions or ideals are cast aside with their denotation as 'dinosaurs'. The act of political witnessing within this usage is clear; rather than constituting a means by which humans live with the environmental heritage, this is a process of distinction. Through association and allusion, the 'natural' is dismissed in the assumption that human society represents progress. Language casts us as political witnesses as it reiterates the established values and ideals of our contemporary world.

The appearance of these modern dinosaurs as a vision of the anachronistic is confirmed with the figurative usage of 'dinosaure' within the French language and synonyms which highlight the outdated character associated with the term. The same usage can be seen within German as 'dinosaurier' or 'saurier' is used to characterise the antiquated or the backward-looking. Indeed, the epithet of 'dinosaurier' is associated with the auslaufmodell (outdated/discontinued), the überbleibsel (remains) or the relikt (remnant/relict). The terms are used to describe those who the rest of society or the 'modern age' has accelerated away from and disregarded. The usage of 'dinosaurier' as an adjective within German begins within the mid-twentieth century as the term provides a cutting indictment of those who do not adapt to the 'real world'. From the mundane to the world of high politics, the accusation of being a 'dinosaur' carries with it a condemnation of lacking the appropriate skills or awareness to exist within contemporary society. From environmental concerns to survival in the corporate world, the engagement with this mode of representation through reference to natural history is highly flexible (see Anon 2016b; Ettel, Seibel and Zschäpitz 2016).

The presence of 'dinosaurs' in these sectors is used to highlight the dangers of such old-fashioned concepts to be allowed to continue. Similar concerns are expressed across other European languages. Whether it is warning that individuals, groups or corporations will come to an abrupt end 'como los dinosaurios' in Spain or 'come i dinosaur' in Italy, the objective behind this reference remains the same. The employment of these allusions serves to direct and protect the shape of society beyond the present, ensuring that the character of modernity is continued with the colonisation of the future (after Giddens 1990). In effect, the emergence of this particular type of discourse perpetuates the divide which has been regarded as characteristic of the modern age (Latour 1993). This is the intangible heritage of natural history, located in the terms, phrases and associations which have been developed within human society. In these references, Nature is cast as a failure and culture is made distinct from its 'natural' associations. Humans remove themselves from the non-human not solely through the ability to use language but through the content of that language. Natural history is used to measure the progress and advancement of our own time and not regarded as part of our world.

This process of measuring, both literal and figural, has been part of the incorporation of scientific terms into the wider vocabulary from the nineteenth century. Both geological terms and the names of extinct animals serve as points of comparison. For example, dinosaur, brontosaurus, diplodocus, tyrannosaurus and mammoth have all become used to describe excessive dimensions and scale over the past century as if drawing upon a collective memory of inefficient, oversized animals which dominate their surroundings. Indeed, from the 1880s and 1890s, advertisements for stores selling garments, foods or sundry items were described with the adjective 'mammoth' as an indication of the extent of the business. 'Mammoth' still possesses a place within advertising culture with shops, clothing and design companies all employing the name for their businesses to communicate scale or strength. Within contemporary political contexts, large, exhausting, potentially unassailable tasks are defined with the title of 'mammoth'. An example of where such descriptions were repeatedly applied to a specific context is the aftermath and negotiations associated with the referendum on continued membership of the European Union in the United Kingdom in June 2016. In such circumstances, where the exact contours of the debate appeared beyond the reach of politicians and economists, the application of 'mammoth' in the popular press appeared to be a useful reminder of the magnitude of the issue (see Inman 2016; Mance and Blitz 2016).

Mammoth, which is itself derived from the Russian mammot, is similarly employed across other European languages within contemporary discourse. Within French, mammouth as an adjective refers to excessive size and is associated with colosse (colossus) and géant (giant). In this manner, 'le mammouth' has become a symbol representing waste, inefficiency and distended institutions. The most famous re-emergence of the extinct animal within contemporary France was in 1997, during a conflict between the French Minister of Education and Socialist Party member, Claude Allègre, and the union of teachers, SNES (Syndicat National des Enseignements de Second Degré/National Union of Secondary Education). Allègre stated in a meeting that, 'il faut dégraisser le mammouth' (we must degrease the mammoth) (Quinio 1997). Depicting a vital part of the state infrastructure as a cumbersome, disproportionate and now non-existent beast casts the education sector as unfit for purpose in the modern world. Such is the power of this analogy that subsequent debates regarding this issue have all revolved around the figure of the 'mammouth'. Indeed, press coverage of the issue regularly resurrects the animal to return to the issue of the efficiency and management of this area of the public sector (see Beyer 2016; Pech 2014).

Indeed, the mammal that once roamed Europe before the Holocene approximately 12,000 years ago appears to have made a reintroduction if the representation within the continent's media is taken into account. Within Germany, mammut is evoked to relay to wider society the size or the unwieldy nature of a particular political, economic or social issue. Indeed, the animal is to be regularly found as a determiner of a compound noun in the assessment of contemporary life: 'Mammut-Reform' (Anon 2006), 'Mammut-Bauprojekt' (Anon 2010) and 'Mammut-Prozess' (Diehl 2011). The association with synonyms such as gewaltig (enormous), kolossal (colossal) or riesig (huge) ensure that the mammoths that are called to mind in such circumstances communicate the intensity and the insurmountable character of tasks.

Dinosaurs and mammoths are prominent examples of this process but further uses of extinct animals or geological periods also provide a framework for cultural expression through metaphor or allusion. For example, excessive periods of time spent awaiting an event or the performance of a responsibility by an individual, corporation or government are couched in terms of 'Ice Ages' or described as 'glacial'. Individuals whose morality, politics or economic outlook appears to be inappropriate to the modern age are cast as 'cavemen', 'Neanderthals' or belonging to the 'Jurassic Age' or else defined as emerging from a 'primordial swamp'. Similarly, longevity in a role, or perhaps long-held establishment within a particular field, is defined as being 'as old as the hills' or compared to the processes of rock formation. These references serve to criticise, rebuke or to undermine current figures or groups as ideas about identity, place and power are negotiated through this heritage. Therefore, across languages and societies in Europe and North America there is an intangible legacy of natural history that functions to direct the present and shape the future. However, this is a future where those divisions between nature and culture are ingrained. Society is formed through the assessment of the human but not with the inclusion of the non-human. Language functions as an act of political witnessing as we testify to our place within the world through the structures present in the references that we employ.

This follows a wider pattern of how nature and natural history are featured in metaphors, allusions and idioms within contemporary society (see Palmatier 1995). This heritage is overwhelmingly used as a means of censoring human behaviour that moves beyond the accepted norms of society. As such, the environment and environmental history features within political, media and public discourse as a negative, drawing further distinction between the human and the non-human (after Kovecses 2002: 153). For instance, to describe an act of excess, whether violence, greed, aggression, pride or timidity in English, French or German, recourse would be made to the natural world and natural history:

tiger / tigre / tiger
whale / baleine / wal
lion / lion / löwe
bear / ours / Bär
mouse / souris / maus

Whether derived from common Latin roots, original native terms, loan words from other languages or scientific names developed from the study of the natural world which emerged from the eighteenth century, these monikers have served as a resource for contemporary western human society to represent its most proscribed behaviour. To act 'like an animal' is to reflect the basest forms of existence and to be regarded as a pariah within society. This division between human and non-human in language is a modern, western phenomenon and derived from the cultural responses formed through capitalism, industrialisation and urbanism (Gross and Vallely 2012). This is largely drawn from the premise that animal life possesses an inferior understanding of being or existence, trapped as it is between its instinctive drives and its environment (after Heidegger 1995: 248). Therefore, animal references and wider associations with natural history, reflect both a repulsion and attraction where the excess represented with this environmental heritage serves as a troubling yet beguiling presence in human society (see Freud 1913). In this manner, it is forever entailed to be the object of fascination as society addresses the tensions it possesses towards the non-human elements of the world. Language, through reference, metaphor and idiom, is reflective of the act of political witnessing that we undertake and the testimonies that are obtained through the separation of nature and culture. However, within this division and the language employed to define our world, the consideration of the 'democracy of things', where the human and the non-human are part of a wider society, can also be formed. The employment of terms or phrases which reflect an awareness of natural heritage can act as a point of connection. Whilst the contemporary references and metaphors cast nature as a negative element, there is a potential to use language to construct a wider, inclusive society by regarding the uses of such allusions as indicative of a world held in common rather than one perceived by the human alone. As such, pronouns become inclusive as there emerges a far broader and far more nuanced assessment of society when 'we' is considered. A natural heritage can use these points of reference and allusion to ensure a 'democracy of things' emerges.

Conclusion

Through theme parks, heritage sites, television, film and language, the role and function of natural heritage can be assessed as locales of witnessing, where society comes to assess itself and others and testify to the significance of these perceptions. These tangible and intangible forms of heritage are significant as they represent an engagement with natural history away from the confines of state museums or national parks. Within each of these locales, individuals, groups and communities interact with the representation of the past to form a sense of identity, place and politics. This process is fundamental to the understanding of our world as it defines how we relate ourselves to the past, the present and potentially the future. It is through these places, whether through physical engagement or media representation, that we learn how to be 'modern'; that is, to bear witness to the separation of the human and the non-human within society. In the moral witnessing that can be noted within the representation of natural history on television, the exercising of ethical positions is highly significant. Indeed, it is through such depictions of natural heritage and environmental history that wider issues regarding climate change or ecological damage can be raised. Nevertheless, the way in which such a moral consensus is formed can serve to undermine the objectives of these highly popular programmes. The concern and acknowledgement that is evoked through such representations can serve to reiterate widely-held perceptions but not alter wider actions, ensuring that the relationships formed between humans and non-humans do not alter. Similarly, the appearance of natural history within language also acts to distinguish elements within society. Through the emergence of scientific discourse of the nineteenth century within the popular vernacular, language has formed a shared memory of the natural world across Europe and North America. This is based upon the premise of the environment serving as a symbolic resource for emphasising and critiquing the excess within human society. Language, whether through metaphor, allusion or reference, forms a means of political witnessing where the modern distinction between the human and the non-human is reiterated and maintained. However, these sites can also form a means to establish an alternative mode of engagement with the natural world that sees shared matters of concern and a wider concept of society. In locations such as dinosaur parks, whilst moral witnessing may not be present, we can observe time, place, extinction, fragility and the recognition of a social world formed from an understanding that we do not impact upon our environment – we are part of that environment.

6 Conclusion

Introduction

From the earliest collectors' cabinets displaying fossils, specimens and curios of natural history in the sixteenth century to the interactive and immersive displays of contemporary museums and the conservation areas protected from the incursion of development, we have made ourselves modern through association with our environmental heritage. This establishment of 'modernity' has at its core a distinction between nature and culture, the separation of the human and the non-human. This division has enabled humanity to regard the environment as a resource, removed from society as an altogether different entity upon which human aspiration, value and ideology are placed. It in within this natural heritage that successive generations have located intellectual, class, gender, ethnic, economic, cultural and political distinction within the palaeontological, geological and climatological history of the Earth. Whilst the environment has been physically mined for its properties, it has also served as a symbolic reserve from which we have extracted a perspective on the past, present and future. This outlook is framed by a sense of control and operation, where the forces of nature are brought under cultural dominion. Such an assumption of authority has brought humanity an ability to classify and organise the operation of 'nature'; a mode of being which has generated a particular sense of place with regard to the environmental heritage (after Latour 1993). Humans have made sense of their surroundings through this awareness, institutions have been organised, legislation has been formed, societies have been structured and even our language reflects these changes. However, as the threat of environmental catastrophe brought about by rapid climatic change threatens the global society, the consideration that this distinction between human and the non-human has been the cause of such damage has been forwarded by scholars (Morton 2013). This assessment is significant as it raises questions regarding our ability to develop solutions to our problems, as any alternative mode of engagement with the natural world is already undertaken in the context of this division between nature and culture. In this manner, 'nature' arrives to us as an object that is already understood (after Althusser 1971: 126).

Towards a critical natural heritage

Natural heritage forms a means by which these well-held notions of place, politics and identity can be challenged as alternative relationships to the environment can be developed through this engagement (after Harrison 2015). By establishing humanity and nature as equal parts of an inclusive notion of society, where the human and the non-human participate in the creation of meanings and values, the redefinition of our modern world can be undertaken. We return to nature in this assessment, not as a nostalgic vision of a pristine wilderness, nor as some object upon which to reflect and contemplate, but as a means of making a new world. Such an undertaking faces considerable challenges. The way in which the distinction between nature and culture have become so embedded within western society is so extensive that it has taken on the appearance of a 'natural' division. Nevertheless, through the critical assessment of the representation of natural heritage through a variety of forms, an insight into the function and potential of a relationship between humans and non-humans can be developed. This would serve as a basis for reimagining the world, as we can conceive of a society that includes all elements of our planet. Rather than assume human predominance, we begin to think of heritage beyond the human. This is not a dereliction of responsibility, where asserting the agency or involvement of non-human objects within society removes the culpability or requirement of humans to address environmental concerns. Rather, it strengthens these associations and enables the emergence of an alternative basis for a different type of society where we speak of things held in common rather than what distinguishes humans from the world.

To regard this shared perspective on matters of concern and to form this vision of an inclusive society we can assume the role of the witness. The significance of this identity is its involvement and engagement with the wider world. Instead of acting as a controlling or defining element, humans are required to observe and to testify as to the significance of the world around them. It is a testimony made in common with the elements of the wider environmental heritage. To understand the effect of these acts of witnessing that take place with regard to natural history, we can consider the wider ethical, governmental and public implications of these activities. As the tangible heritage sites of museums, national parks, conservation areas, theme parks or the intangible heritage of media representation, memory and language are examined, acts of witnessing are performed that affirm, alter or disrupt the operation of power and authority within contemporary society (after Smith 2006; Harrison 2013). By assessing the role of natural heritage in forming moral, political or social witnesses, we can understand the function of this resource for individuals and communities. In this manner, engaging with this past forms a means of changing the conditions of contemporary society and the shape of things to come. It can be observed that museums and national parks, which have undertaken important work in recent years on issues of education and conservation, have established concepts of personal and collective responsibility towards the environment; that legislation and language affirm the structures of authority and the ideological divisions within our communities; and that sites where we engage with social identities beyond the human provide the basis of forming an alternative society.

Over the last few centuries, western society has remade itself and remodelled the past, present and future through natural heritage. As the threats posed to humanity from environmental change continue to materialise, the division between nature and culture which has marked the modern world cannot serve to address these crises when it is the driving force of our current predicament. Therefore, a study of natural heritage can provide an alternative model of this relationship. This is not achieved by some romantic act of reverie where a point in the past is elevated as it is regarded as representing a moment of equilibrium. Rather, as natural history has always represented a shifting set of values and ideals, such equilibrium can be a product of negotiation and engagement in the present. Through this heritage, we can regard the relationship between the human and the non-human as a dialogue, a process of deciding issues held in common. Sites where acts of social witnessing take place enable this process to occur as we testify through these places that we exist alongside others as part of the same society. It is this democratisation of nature and the environment that Latour (2004a) forwards as a critical and constructive political ecology for the twenty-first century. For natural heritage, this can be formed through acknowledging the formation of place, politics and identity as part of witnessing and testifying within these locales. In this manner, a new concept of the environment can be formed which can redefine our relationship to the 'natural' as an element that we are part of rather than an object of our observation and control. Museums, national parks, conservation areas, language, memory, theme parks and media representations all possess the ability to act as spaces where the human and the non-human coexist.

Coexistence: forming a new model of society

One of the features of the vibrant area of political and philosophical environmental studies over the past two decades has been the emergence of a sustained critique of the predominant western notions regarding the relationship between humans and nature (Doherty and Doyle 2008). This has resulted in a wave of scholarship which has focused on redefining contemporary society along an ecological perspective where humans are regarded as one element amongst a wider array of life on Earth (see Lamb 1996; Winter 1996). In part, this stems from the rise of environmentalism in western society from the 1960s as a growing concern for the impact of pollution brought greater focus on the ideological frames used to perceive 'nature' (see Carson 1962). However, a distinctive turn towards 'things' is discernible across the literature of political ecology from the early 1990s as scholars forwarded concepts of coexistence as a means of forming less damaging engagements between humans and the natural world (see Cronon 1997). This follows a wider pattern across the social sciences which has focused upon the intersection between the objects and agents within society rather than the divisions or categories that have been placed upon these issues of concern (Gell 1998; Latour 1993). Within environmental studies, this approach has manifested itself in establishing a recognition that previous conservation agendas which sought to protect and preserve the natural world are compromised by their association with a culture that regards 'nature' as an object removed (see Harvey 1996). In the interpretative space that has opened up with this development, a number of scholars have suggested new models of society to move beyond the human (see Viveiros de Castro 2014).

This movement has provided a means of understanding a 'contested nature', composed of varying agendas, attitudes and ideals, as humans attempt to comprehend their world and address the threats of environmental change (after Macnaghten and Urry 1998). The critique of 'nature' as possessing a quality beyond human interpretation or a fundamental category that cannot be altered has been significant in realising the created and collaborative aspects of the environment. This is more than the reductive argument of realism or constructivism which has dominated environmental science studies in the past (Arias-Maldonado 2012). In these assessments, realism identifies nature as possessing a tangible, irreducible character, whilst constructivism emphasises the socially and culturally specific discourses that have formed our perception of nature (Soulé and Lease 1995). However, such frameworks do not encompass the varied and multiform means by which individuals and communities engage with natural heritage (see Gottlieb 1997). Indeed, they also perpetuate the division between nature and culture without advancing the debates on how humans and non-humans co-exist (Inglis, Bone and Wilkie 2005). The ontological issues raised by these debates have been addressed by a number of studies within the field of anthropology. One of the most prominent of these works is the assessment of non-western, indigenous attitudes towards nature by Descola (1994; 2013). In this examination, a different basis of society is forwarded for living with nature as the association with the environment within global societies provides points of reflection on practice, identity and ideology. As such, Descola (1994) highlights the ways in which we live with nature, through analogism, animism, naturalism and totemism, not apart from the environment. A similar line of inquiry has been pursued by Kohn (2013), who has emphasised the notion of living with nature by asserting the agency of the environment. Essentially, this demonstrates how the qualities we might regard as uniquely human are exhibited by the natural world (Kohn 2015).

Ingold (2000) has also followed this area of study with an assessment of how the divisions between biology and culture have been imposed by successive generations of human societies rather than being derived from some innate character. This 'ontological turn' within anthropology has revealed the potential of moving beyond the human and addressing society in context with the environment (Henare, Holbraad and Wastell 2007). It has also been conducted with a strong focus on the future and the way in which society can respond to the threats to its existence. This orientation towards the shape of things to come has marked this development within anthropology and it is also reflected across the wider social sciences which have followed this ontological turn (see Latour 2014). Such an agenda is a response to environmental concerns but also the emergence of the definition of contemporary society as living within the 'Anthropocene' (Waters et al. 2016). This designation has been used to describe how, as the climate and geology have been so radically altered by human activity in recent history, the period from the mid-eighteenth century to the present would be accurately termed as a departure from the Holocene. With this development, the relationship between past, present and future is brought under greater scrutiny as humanity's defining role in the contemporary era can be reassessed in the light of historical trends or potential developments (after Dawdy 2009). Therefore, the recognition of the 'Anthropocene' presents scholars of heritage studies with a dilemma (see Solli et al. 2011). Whilst the human involvement in the changes witnessed across the planet are evident, the notion of forming universal values from such a recognition would go against the established critique of essentialist definitions of 'heritage' (after Smith 2006). In response, Harrison (2013; 2015) has regarded the function of an 'ontological politics of heritage' as both a pluralist and a creative endeavour which seeks to establish different pasts and different futures (Harrison 2015: 28). As the 'Anthropocene' is used to define and cast uncertainty in our own era, heritage studies can recognise the coexistence of humans and non-humans in building alternative societies. Rather than 'colonising the future', this is a process of 'negotiating a future' through the past and present.

This enmeshing of society and the formation of alterity and alternatives has been examined by Haraway (1991; 2003; 2008) as a means of adjusting to rapidly changing local and global conditions. Indeed, in Haraway's (2016) assessment, the term 'Anthropocene' is dismissed as inaccurately representing the complex and multiform ways in which humans and non-humans have interacted. In its place, Haraway (2016) suggests the 'Chthulucene', referring to the myriad of ways in which we imagine human lives with non-human lives. Haraway (2015) suggests that this term forms the basis of recognising the complex composition of the environment, as it:

> . . . entangles myriad temporalities and spatialities and myriad intra-active entities-in-assemblages – including the more-than-human, other-than-human, inhuman, and human-as-humus (Haraway 2015: 161).

It is the sense of life lived with others that Haraway (2016) defines as key in making new societies beyond the conflicting and constraining divisions that humans in the modern era have made for themselves and consider to be 'natural'. Heritage studies is crucial in this endeavour as it serves as a source of new relations, new connections, alternative pasts and hopeful futures. The latter is significant here as we move beyond the critique (not the critical) to form the constructive, to build new societies and to consider coexistence (after Haraway 2015). It is within this 'Chthulucene' that a natural heritage studies can emerge which regards the past life of humans and non-humans as a present concern and a common matter for the future. Time, place, identity and politics intersect these processes as we move beyond the human to consider a social world that includes all aspects of life. Therefore, the Bergsonian (Deleuze 1988) concepts of la durée, the élan vital and memory provide guides to engaging with natural heritage. Bergson (1908; 1910) rejected the strict formulation of natural history which separated time, space and life in the moment of creation and engagement. By locating the dynamism of life in the past, present and future across all forms of existence, the connections and coexistence within natural heritage can be witnessed and new modes of the social can be formed (Bergson 1908: 24).

Witnessing natural heritage

By serving as witnesses to natural heritage we can engage, affirm or disrupt the established structures within society. To be a witness does not entail some passive observation of the world without providing a response. A witness will actively engage in what is observed and bear testimony as to its significance. The witness is there to acknowledge and be responsible for the event. It is not a matter of standing in judgement but a means of standing beside the object or event to record this process. It is the notion of responsibility and testimony that ensures that the act of witnessing is a constructive performance. The role of the witness within the study of natural history is used within this assessment to assess the responses formed through the tangible and intangible representations of this past. The witnesses formed through the sites of natural history across contemporary western society are not automatons, as locales do not structure or frame individual and collective experience so thoroughly. However, within these areas there are distinct 'senses of place' that serve as stages upon which the act of witnessing can take place. The acts of moral, political or social witnessing defined in this work place people in the context of their environmental history in particular ways, whether it is acknowledging the political structures present within representations which enact the ideology of our own era upon the past, the recognition of moral claims made upon us by our relationship to the natural world, or by realising the social context within which the human and non-human coexist. The figure of the witness, rather than being a fabricated category or rhetorical device deployed for the sake of argument, has been employed for its established place within western culture. Indeed, as humans are faced with ever-increasing threats to our continued existence on the planet, it is the witness who can act as an important bulwark against the tide (see Bird Rose 2004).

The character of the witness has been discussed by Haraway (1997) as a vital component within a democratic system but one which has to be considered carefully. Haraway's (1997: 26) study of the emergence of scientific practice and discourse in the eighteenth century assessed the gentlemanly character of the 'modest witness' who noted the process of experimentation and then provided an objective account of what had transpired. However, this apparently neutral assessment was loaded with assumptions regarding gender, class, ethnicity and politics. Haraway (1997: 15) rehabilitates this figure of the 'modest witness' as a means to examine the structures of power within historical and contemporary scientific discourse. As a role which is formed through a nexus of social, political and cultural values, the 'modest witness' in this guise is a figure of responsibility, not a pretence of objectivity (Haraway 1997: 155). The witness engages in a connected and collaborative process with human and non-human others to maintain, alter and enhance the world around them (Haraway 1997: 269). The witness is, therefore, crucial in shaping society:

> Witnessing is seeing; attesting; standing publicly accountable for, and psychically vulnerable to, one's visions and representations. Witnessing is a collective, limited practice that depends on the constructed and never finished credibility of those who do it, all of whom are mortal, fallible, and fraught with the consequences of unconscious and disowned desires and fears (Haraway 1997: 267).

The responsibility of the witness is to ensure that the meanings that are derived from their observation, albeit partial and subjective, are brought forth to engage and shape our sense of place. This sense of obligation has been explored by Bird Rose (2004: 213), who has regarded the role of the witness as the most important element in addressing climate change, pollution and environmental deterioration. The witness in this perspective is attentive, connected and responsive to the world as they engage with the lives of others, both human and non-human (Bird Rose 2004: 213). The witness is key to holding society to account:

> Those who witness to the loss of place, especially when place is lost to the colossus of development, take up a moral burden. They break up monologue and sustain a moral engagement with the past in the present that gives voice, presence, and power to that which has been lost, abandoned, or destroyed. The burden of witnessing to ecological loss brings Nature into a moral community, implicitly asserting in the context of place the injunction . . . not only to speak truthfully, but to remain true to the place. In remaining true to place, one remains (or seeks to remain) true to non-human living things, ecosystem, and processes of resilience (Bird Rose 2004: 51).

In this manner, the role of the witness is not to preserve a legacy but to remake the world by ensuring voices are represented and acknowledged for the past, present and future (after Margalit 2002). It is not the accuracy or the completeness of the witness's testimony that is significant, it is the act of attentiveness and a commitment to responsibility that underlines the role of the witness (Hatley 2000). Therefore, to consider the role of the witness with regard to natural heritage is to assess an act of 'opening' to wider contexts. However, rather than invest the witness with an essential, reforming, progressive zeal, we can acknowledge the role of the witness in affirming the norms and values within society as much as challenging its stability (after Haraway 1997: 15). The various functions of the role are characterised here as forming acts of moral, political and social witnessing; these are used to define how museums, national parks, legislation, heritage trails, tourist parks, media representation and language all serve to provide places where witnessing takes on a particular shape and form (after Bird Rose 2004: 51). Within these tangible or intangible locales, performances of witnessing natural heritage define our society; whether it is through the realisation of a moral truth, a political structure or a social context, natural heritage sustains and remakes us in this moment of engagement. Within this assessment, it is the role of social witnessing which has been characterised as a force for acknowledgement, responsibility and change within contemporary western society. Through an act of witnessing individuals and groups within a wider context of other humans and non-humans, a new concept of the social can emerge (Latour 2004a). This is the value of a critical natural heritage studies; it can serve as a tool of critique to expose the underlying assumptions and relations of power within what might appear to be 'natural'. Conversely, it also acts as a means of dialogue and debate between people, objects, animals, spaces and places which serves to build alternative pasts, presents and futures (after Harrison 2015).

Epilogue

Through a study of museums, parks, legal structures, guides, media representations and language, the role of natural heritage in the formation of the modern world can be assessed. Concepts of the 'natural' have been essential in establishing the political, social, cultural and economic structures that have defined individual, community and state identities over the past four centuries. Indeed, natural heritage has made the modern world. However, as the threats posed by industrialisation, urbanisation and late capitalism threaten the survival of the planet in a manner never experienced before within human history, this modern relationship with nature has to be rethought. This can be done through a critical assessment of natural heritage within society. The display of fossilised animals, prehistoric mammals, ancient landscapes, geological formations, media representations or dinosaur models provides a point of connection for contemporary society to consider time, place and identity. As such, this engagement with our environmental legacy possesses a dissonant, perhaps revolutionary, perspective, as it serves as a challenge to the structures of contemporary life by establishing humanity's place as one element in a shared environment. Natural heritage forms a vital point for demonstrating the relationship between humans and non-humans, which in a society that has divided and distanced nature and culture is a vital resource. In part, this is perhaps a continuance of humanity's repeated return to nature within the modern era to define itself in the face of change and development. Regardless of this appearance, rather than constituting another example of the processes of modernity, this is a recognition of a shared environment where humans are not defined by their dominance but characterised by their coexistence. This is what is revealed in the act of social witnessing – the points of connection and collaboration that we share with ourselves and others within the environment. This is the orientation of the social witness, formed from dialogue between what is, what has been and what is still yet to come. The social witness to natural heritage bridges these ontological divisions to regard us all as existing with one another, within time and within place as a means of considering an alternative society. In this manner, we bear witness to the past, present and future in the hope of a better world.

Bibliography

6ter. 2016. La vraie vie des dinosaurs. Available at: www.6ter.fr/emission-xplora/15-12-2012-la_vraie_vie_des_dinosaures-2147513814.html [Accessed 1 August 2016].

Acres, B. and Paul, R.W. 1896. *Rough Sea at Dover*. s.n. [Film]

Agassiz, L. 1859. *An Essay on Classification by Louis Agassiz*. London: Longman, Brown, Green, Longmans, & Roberts.

Agassiz, L. 1866. *The Structure of Animal Life: Six Lectures*. London: Sampson Low, Son and Marston.

Alberch P. 1993. Museums, collections and biodiversity inventories. *Trends in Ecology & Evolution*, 8: 372–375.

Alberti, S.M.M. (ed.) 2011. *The Afterlives of Animals: A Museum Menagerie*. Charlottesville and London: University of Virginia Press.

Albright, M. 1998. *Earth Day 1998: Global Problems and Global Solutions 1997–2001*. Available at: state.gov/www/statements/1998/980421.html [Accessed 21 July 2016].

Althusser, L. 1971. *Lenin and Philosophy and Other Essays* [Trans. B. Brewster]. London: New Left Books.

American Museum of Natural History. 1941. *Seventy-Third Annual Report*. New York: The American Museum of Natural History.

American Museum of Natural History. 1948. *Eightieth Annual Report*. New York: The American Museum of Natural History.

American Museum of Natural History. 1953. *General guide to the exhibition halls of the American Museum of Natural History*. New York: American Museum of Natural History.

American Museum of Natural History. 1984. *Ancestors: Four Million Years of Humanity. American Museum of Natural History, April 13 – September 9, 1984*. New York: American Museum of Natural History.

American Museum of Natural History. 2016a. Anne and Bernard Spitzer Hall of Human Origins. Available at: www.amnh.org/exhibitions/permanent-exhibitions/human-origins-and-cultural-halls/anne-and-bernard-spitzer-hall-of-human-origins/understanding-our-past [Accessed 20 May 2016].

American Museum of Natural History. 2016b. Harry Frank Guggenheim Hall of Minerals. Available at: www.amnh.org/exhibitions/permanent-exhibitions/earth-and-planetary-sciences-halls/harry-frank-guggenheim-hall-of-minerals [Accessed 20 May 2016].

American Museum of Natural History. 2016c. Arthur Ross Hall of Meteorites. Available at: www.amnh.org/exhibitions/permanent-exhibitions/earth-and-planetary-sciences-halls/arthur-ross-hall-of-meteorites/meteorites [Accessed 20 May 2016].

Anderson, A. 1997. *Media, Culture and the Environment*. London: UCL Press.

Anderson, A. 2014. *Media, Environment and the Network Society*. Basingstoke: Palgrave Macmillan.

Anderson, B. 1983. *Imagined Communities: Reflections on the Origin and Spread of Nationalism*. London: Verso.

Anon. 1790. *A Companion to the Museum*. London: s.n.

Anon. 1829. Some Remarks on Natural History, as a Means of Education. *Magazine of Natural History*, 1: 10–14.

Anon. 1840. *Charter and supplement to charter and by-laws of the Philadelphia Museum Company*. Philadelphia: A. Waldie.

Anon. 1841. *Provincial Ordinances of Lower Canada, vol. 16*. Quebec: John Charlton Fisher and William Kemble.

Anon. 1842. *Synopsis of the Contents of Gesner's Museum of Natural History at Saint John*. Saint John: Henry Chubb.

Anon. 1845. Angelegenheiten des Vereins. Jahreshefte des Vereins für vaterländische. *Naturkunde*, 1: 1–8.

Anon. 1846. *Abhandlungen aus dem Gebiete der Naturwissenschaften herausgegeben von dem naturwissenschaftlicher Verein in Hamburg*. Hamburg: Agentur des Rauhen Hauses.

Anon. 1851a. *Notes and sketches of lessons on subjects connected with the Great exhibition*. London: Society for Promoting Christian Knowledge.

Anon. 1851b. *Reports by the Juries on the Subjects in the Thirty Classes into Which the Exhibition was Divided*. London: William Clowes and Sons.

Anon. 1852a. *The Industry of Nations, as Exemplified in the Great Exhibition of 1851*. London: Society for Promoting Christian Knowledge.

Anon. 1852b. *Mittheilungen über neue Erwerbungen des naturhistorischen Museums in Hamburg*. Hamburg: s.n.

Anon. 1861. *Report of the Trustees of the Museum of Comparative Zoology*. Boston: William White.

Anon. 1864–1865. *Memoirs of the Museum of Comparative Zoology at Harvard College, Vol. 1*. Cambridge, MA: Harvard University Press.

Anon. 1866. *Report of the Sheffield Scientific School of Yale University*. New Haven, CT: E. Hayes.

Anon. 1870. *First Annual Report of the American Museum of Natural History*. New York: Major & Knapp.

Anon. 1871. *An account of the organization and progress of the Museum of comparative zoology*. Cambridge, MA: Welch, Bigelow and Company.

Anon. 1872. *Natur und Museum*. Senckenbergische Naturforschende Gesellschaft: s.n.

Anon. 1883. *Annual report of the Board of Trustees of the Public Museum of the City of Milwaukee*. Milwaukee, WI: Milwaukee Public Museum.

Anon. 1894. *An historical and descriptive account of the Field Columbian Museum*. Chicago, IL: Field Columbian Museum.

Anon. 1901. *An Account of the Horniman Free Museum and the Recreation Grounds, Forest Hill*. London: Horniman Free Museum.

Anon. 1904. *A Guide to the Collections of The Horniman Museum and Library*. London: Horniman Museum.

Anon. 1912. *Guide for the use of visitors to The Horniman Museum and Library*. London: Odhams Ltd.

Anon. 1915. *Bericht über das Zoologische Museum in Berlin im Rechnungsjahr 1914. Milleilungen as dem Zoologischen Museum in Berlin*. Berlin: In Kommission bei R. Friedlander & Sohn.

Anon. 1916a. One who was there. *University Magazine,* 15(4) (McGill University): 458–469.

Anon. 1916b. Anti Preparedness Committee Launches Huge Model of Armored Dinosaur as satire on Military Preparedness. *The American Socialist*, April 15, 2.

Anon. 1916c. An animal of extinction. *The Survey*, 36: 165.

Anon. 1916d. Notes and Comments. *The Naturalist: a monthly illustrated journal of natural history for the North of England,* 710: 81–91.

Anon. 1918. Museumsbericht uber die Jahre 1915 und 1916. *Natur und Museum,* 47: 36–50.

Anon. 1927. *Führer durch das Museum für Naturkunde und Vorgeschichte in Dessau.* Dessau: Museum f. Naturkunde und Vorgeschichte in Dessau.

Anon. 1942. Birds of the Pacific War Zone. *The Living Museum,* 4(5): 39.

Anon. 1960. The Detection of Nuclear Tests. *Nuclear Information,* 2(6): 1–6.

Anon. 1964. *Wilderness preservation system. Hearing before the Subcommittee on Public Lands of the Committee on Interior and Insular Affairs, House of Representatives, Eighty-Eighth Congress, second session.* Washington, DC: U.S. Government Printing Office.

Anon. 1972. *The American Museum of Natural History: an introduction.* New York: The American Museum of Natural History.

Anon. 1974. Museum of Natural History – Institutional Racism. *Black News,* 2(19): 8–9.

Anon. 1977. *Human biology: an exhibition of ourselves.* London: British Museum.

Anon. 1979. *Dinosaurs and their living relatives.* London: British Museum.

Anon. 1985. Aboriginal Gallery. In *Australian Museum Annual Report, 1984/85* (27–28). Sydney: Australian Museum.

Anon. 1988. *Tropical Rainforests: A Disappearing Treasure.* Washington, DC: Traveling Exhibition Service, Smithsonian Institution.

Anon. 1998. Dinosaur doorman: a diplodocus will greet visitors to Carnegie Museum. *Pittsburgh Post-Gazette*, April 24.

Anon. 1999a. *The Kyoto Protocol: is the Clinton-Gore administration selling out Americans? Parts I–VI: hearing before the Subcommittee on National Economic Growth, Natural Resources, and Regulatory Affairs of the Committee on Government Reform and Oversight, House of Representatives, One Hundred Fifth Congress, second session, April 23; May 19 and 20; June 24; July 15; and September 16, 1998.* Washington, DC: General Printing Office.

Anon. 1999b. Editorial: Diplodocus Oaklandus. *Pittsburgh Post-Gazette*, July 9.

Anon. 2003. Time Marches On. *Pittsburgh Post-Gazette*, November 2.

Anon. 2005. *Caves win 'natural wonder' vote.* Available at: news.bbc.co.uk/1/hi/wales/mid/4735935.stm. Accessed 20 August 2016 [Accessed 14 June 2016].

Anon. 2006. *Mammoth Reform: Coalition to plan health tax for all.* Available at: www.spiegel.de/politik/deutschland/mammut-reform-koalition-soll-gesundheitssteuer-fuer-alle-planen-a-424161.html [Accessed 12 July 2016].

Anon. 2010. *Mammoth Construction Project: China Settles 330,000 People.* Available at: www.spiegel.de/politik/ausland/mammut-bauprojekt-china-siedelt-330-000-menschen-um-a-711481.html [Accessed 19 August 2016].

Anon. 2014. Vingt ans que la galerie ramène sa science. *Le Parisien*, 19 September.

Anon. 2015. Campaign to Save Dippy Diplodocus. *The Herald*, January 30.

Anon. 2016a. 'Corbyn was never taken very seriously'. Labour must ditch its dangerous dinosaur. *Daily Express*, July 24.

Anon. 2016b. *RWE bringt Tochter Innogy an die Börse. Die Welt, October 10.* Available at: www.welt.de/regionales/nrw/article158605865/RWE-bringt-Tochter-Innogy-an-die-Boerse.html [Accessed 12 October 2016].

Appadurai, A. 1995. The Production of Locality. In R. Fardon (Ed.), *Counterworks: Managing the Diversity of Knowledge* (204–225). London: Routledge.

Arias-Maldonado, M. 2012. *Real Green: Sustainability After the End of Nature.* Farnham: Ashgate.

Arnold, K. 2006. *Cabinets for the Curious: Looking Back at Early English Museums.* Aldershot: Ashgate.

ARTE. 2011. *Wildes Deutschland.* ARTE Television [Documentary].

ARTE. 2015. *Planet Deutschland – 300 Millionen Jahre.* ARTE Television [Documentary].

ARTE. 2016a. *Die Odyssee der einsamen Wölfe.* ARTE Television [Documentary].

ARTE. 2016b. *Frauen und Ozeane.* ARTE Television [Documentary].

Article L331–1. 2016. *Code de l'environnement.* Available at: www.legifrance.gouv.fr/affichCodeArticle.do?cidTexte=LEGITEXT000006074220&idArticle=LEGIARTI000006833521&dateTexte=&categorieLien=cid [Accessed 13 July 2016].

Asma, S.T. 2003. *Stuffed Animals and Pickled Heads: The Culture and Evolution of Natural History Museums.* Oxford: Oxford University Press.

Aspel, J. 2015. *Introducing Peace Museums.* Abingdon and New York: Routledge.

Augé, M. 1995. *Non-Places: An Introduction to Supermodernity.* London: Verso.

Australian Heritage Commission Act. 1975. *National Estate 4(1).* Canberra: Australian Government.

Avent, C.J. 1990. Rain Forest Travels to San Diego. *Los Angeles Times,* June 30.

Baker, M. and Gordon, J.E. 2012. Unconformities, schisms and sutures – geology and mythology in Scotland. In E. Ellsworth and J. Cruse (Eds.), *Making the Geologic Now* (163–169). New York: Punctum Books.

Bakker, R.T. 1975. Dinosaur Renaissance. *Scientific American,* 232(4): 58–78.

Barthel-Bouchier, D. (Ed.). 2016. *Cultural Heritage and the Challenge of Sustainability.* Abingdon and New York: Routledge.

Bassett, K., Griffiths, R. and Smith, I. 2002. Cultural industries, cultural clusters and the city: the example of natural history film-making in Bristol. *Geoforum,* 33(2): 165–177.

Bastmeijer, K. (Ed.). 2016. *Wilderness Protection in Europe. The Role of International, European and National Law.* Cambridge: Cambridge University Press.

Bates, M. 1950. *The nature of natural history.* New York: Scribner.

Bather, F.A. 1914. Patriotism in the museum. *Museums Journal,* 14: 249–253.

Bather, F.A. 1915. Museums and the War. *Museums Journal,* 15: 2–10.

Batz Jr., B. 1999. The skeleton in Dippy's closet: He's more than one dinosaur. *Pittsburgh Post-Gazette,* July 4.

Baudrillard, J. 1994. *Simulacra and Simulation.* Ann Arbor, MI: University of Michigan Press.

Bauer, A. 2009. The Terroir of Culture: Long-term History, Heritage Preservation, and the Specificities of Place. *Heritage Management,* 2(1): 81–103.

BBC. 2011. *Natural History on the BBC.* Available at: www.bbc.co.uk/pressoffice/pressreleases/stories/2011/07_july/08/bbchistory.shtml [Accessed 12 August 2016].

Beaty Biodiversity Museum. 2015. *Home.* Available at: http://beatymuseum.ubc.ca/connect/about/vision-mission/ [Accessed 7 August 2016].

Beck, J.M. 1917. *The war and humanity: a further discussion of the ethics of the world war and the attitude and duty of the United States.* New York: G. P. Putnam's Sons.

Beck, U. 1992. *Risk Society: Towards a New Modernity*. London: Sage.

Beck, U. 2006. *Cosmopolitan Vision*. Cambridge and Malden, MA: Polity.

Bellous, R.E. (Ed.). 1969. An Exhibition of America's Black Heritage. *Los Angeles County Museum of Natural History, History Division Bulletin*, 5: 42–64.

Bennett, T. 1995. *The Birth of the Museum: History, Theory, Politics*. London and New York: Routledge.

Bennett, T. 2004. *Pasts Beyond Memory: Evolution, Museums, Colonialism*. London and New York: Routledge.

Bentham, J. 1823. *An Introduction to the Principles of Morals and Legislation*. London: W. Pickering.

Berger, J. 1972. *Ways of Seeing*. London: BBC Books.

Bergson, H. 1908. *L'Évolution créatrice*. Paris: Félix Alcan.

Bergson, H. 1910. *Time and free will, an essay on the immediate data of consciousness* [Trans. by F.L. Pogson]. London: George Allen and Unwin.

Bernstein, M.A. 1994. *Foregone Conclusions: Against Apocalyptic History*. Berkeley, CA: University of California Press.

Berry, R.J. 1988. Natural history in the twenty-first century. *Archives of Natural History*, 15(1): 1–14.

Beyer, C. 2016. Absentéisme des enseignants: le «mammouth» incapable d'anticiper et de prévenir. *Le Figaro*, 2 March.

Béziat, E. 2016. Faire de la SNCF une société anonyme, la proposition-choc de deux députés. *Le Monde*, 19 October.

Bird Rose, D. 2004. *Reports from a Wild Country: Ethics for Decolonization*. Sydney: University of New South Wales Press.

Blue Planet. 2001. BBC Productions [Documentary].

BNatSchG.1976.*GesetzüberNaturschutzundLandschaftspflege(Bundesnaturschutzgesetz)*. *Bundesgesetzblatt Teil I. Nr. 147 vom 23.12.1976*.

BNatSchG. 2009. *Gesetz über Naturschutz und Landschaftspflege*. Available at: www.bfn.de/fileadmin/MDB/documents/themen/monitoring/BNatSchG.PDF [Accessed 5 September 2016].

Boon T., and Gouyon, J.B. 2015. The origins and practice of science on British television. In M. Conboy, and J. Steel (Eds.), *The Routledge Companion to British Media History* (470–483). London: Routledge.

Booth, E.T. 1901. *Catalogue of the cases of birds in the Dyke Road Museum, Brighton (Third Edition)*. Brighton: King, Thomas and Stack.

Booth, W.J. 2006. *Communities of Memory, On Witness, Identity, and Justice*. Ithaca, NY: Cornell University Press.

Bourcier, N. 2016. Donald Trump, "fulminant dinosaure sexiste" pour ses détracteurs. *Le Monde*, 10 October.

Bousé, D. 2000. *Wildlife Films*. Philadelphia: University of Pennsylvania Press.

Brackx, A. 1980. Man's place in evolution. *Spare Rib*, 97: 16.

Brämer, R. 2007. *Kur Natur – Regeneration durch Wandern Daten, Fakten und Quellen im Detail. Stoffsammlung Kurnatur-Wandern*. Available at: www.wanderforschung. de/files/stoffsammlung-kurnaturwandern-2_1406231522.pdf [Accessed 17 June 2016].

Brämer, R. 2009. Gesunde Natur ist schöne Natur – Wandern erschließt die heilenden Potenziale der natürlichen Umwelt. In: *Bundesamt für Naturschutz (Hrsg) Tagungsdokomentation Naturschutz & Gesundheit*. Bonn: Allianzen für mehr Lebensqualität in Bonn vom 26.-27.05.2009, pp.55–58

Brämer, R. 2010. Heile Welt zu Fuß Empirische Befunde zum spirituellen Charakter von Pilgern und Wandern. Available at: www.wanderforschung.de/files/heile-welt-zu-fuss1265034962.pdf [Accessed 11 June 2016] .

Brecon Beacons. 2016. *Home.* Available at: www.breconbeacons.org/history [Accessed 3 August 2016].

British Museum (Natural History). 1886. *A guide to the exhibition galleries of the department of geology and palaeontology.* London: Harrison & Sons.

Brocx. M. and Semeniuk, V. 2007. Geoheritage and geoconservation – history, definition, scope and scale. *Journal of the Royal Society of Western Australia,* 90: 53–87.

Brown, T. 1840. *The taxidermist's manual, or, The art of collecting, preparing, and preserving objects of natural history. Fifth Edition.* Glasgow: A. Fullarton.

Browne, A.M. 1878. *Practical Taxidermy* (Second Edition). London: L. Upcott Gill.

Browne, M. and Floyd, J. 2006. Hutton trail opens. *Earth Heritage,* 26, 28.

Brunsden, D. 2003. *The Official Guide to the Jurassic Coast: Dorset and East Devon's World Heritage Coast.* Wareham: Coastal Publishing.

Brunsden, D. and Edmonds, R. 2009. The Dorset and East Devon Coast: England's Geomorphological World Heritage Site. In P. Migon (Ed.), *Geomorphological Landscapes of the World* (211–221). Dordrecht: Springer.

Brute Force. 1914. D.W. Griffith (Dir.). General Film Company [Film].

Buchanan, A. and Wicksteed, G.W. 1845. *The Revised Acts and Ordinances of Lower-Canada.* Montreal: S. Derbishire & G. Desbarats.

Bullock, W. 1808. *A companion to the Liverpool museum.* Hull: Printed for the Proprietor.

Bullock, W. 1812. *A companion to Mr. Bullock's London Museum and Pantherion (Twelfth Edition).* London: Printed for the Proprietor.

Bullock, W. 1817. *A Concise and Easy Method of Preserving Objects of Natural History.* London: Printed by the Proprietor.

Burke Museum. 2016. *Washington's First Dinosa*ur. Available at: www.burkemuseum.org/exhibits/washingtons-first-dinosaur [Accessed 17 July 2016].

Burkhardt, Jr., R.W. 1977. *The Spirit of System: Lamarck and Evolutionary Biology.* Cambridge, MA: Harvard University Press.

Calaforra, J.M. and Fernández-Cortés, A. 2006. Geotourism in Spain: resources and environmental management. In R.K. Dowling and D. Newsome (Eds.), *Geotourism: sustainability, impacts and management* (199–220). Amsterdam: Elsevier.

Calgary Zoo. 2016. *Prehistoric Park.* Available at: www.calgaryzoo.com/animals/prehistoric-park [Accessed 1 August 2016].

Campbell, V. 2013. Framing Environmental Risks and Natural Disasters in Factual Entertainment Television. *Environmental Communication,* 8(1): 58–74.

Canadian Museum of Nature. 2015. *Home.* Available at: http://nature.ca/en/home [Accessed 12 December 2015].

Cardo Land. 2016. *Home.* Available at: www.cardoland.com [Accessed 13 June 2016].

Carson, R. 1962. *Silent Spring.* Boston, MA: Houghton Mifflin.

Cato, P.S. and Jones, C. (Eds.). 1991. *Natural History Museums: Directions for Growth.* Texas: Tech University Press.

Catsadorakis, G. 2007. The Conservation of Natural and Cultural Heritage in Europe and the Mediterranean: A Gordian Knot? *International Journal of Heritage Studies,* 13(4–5): 308–320.

Chicone, S.J. and Kissel, R.A. 2014. *Dinosaurs and Dioramas: Creating Natural History Exhibitions.* Walnut Creek, CA: Left Coast Press.

Cleveland Museum of Natural History. 2016a. *Home*. Available at: www.cmnh.org/about-the-museum [Accessed 21 July 2016].

Cleveland Museum of Natural History. 2016b. *Home*. Available at: www.cmnh.org/visit/planetarium-observatory [Accessed 21 July 2016].

CMNH. 2003. *Home*. Available at: www.carnegiemnh.org/online/dinomite/index.htm [Accessed 17 July 2016].

Coates, P. 1998. *Nature: Western Attitudes Since Ancient Times*. Cambridge, England: Polity Press.

Cochrane, J. 2008. Tourism, Partnership and a Bunch of Old Fossils: Management for Tourism at the Jurassic Coast World Heritage Site. *Journal of Heritage Tourism*, 2(3): 156–167.

Conn, S. 1998. *Museums and American Intellectual Life, 1876–1926*. Chicago, IL and London: The University of Chicago Press.

Constitución Española. 1978. Constitución Española. Available at: www.congreso.es/docu/constituciones/1978/1978_cd.pdf [Accessed 13 July 2016].

Constitution of Lithuania. 1992. *Constitution of Lithuania*. Available at: http://www3.lrs.lt/home/Konstitucija/Constitution.htm [Accessed 22 August 2016].

Convention on the Conservation of European Wildlife and Natural Habitats.1979. Convention on the Conservation of European Wildlife and Natural Habitats. Available at: www.coe.int/en/web/conventions/full-list/-/conventions/treaty/104 [Accessed 13 August 2016].

Coombes, A.E. 1994. *Reinventing Africa: Museums, Material Culture and Popular Imagination in Late Victorian and Edwardian England*. New Haven, CT: Yale University Press.

Cooter, R. and Pumfrey, S. 1994. Separate Spheres and Public Places: Reflections on the History of Science Popularization and Science in Popular Culture. *History of Science*, 32(3): 237–267.

Cottle, S. 2004. Producing Nature(s): On the Changing Production Ecology of Natural History TV. *Media, Culture & Society*, 26(1): 81–101.

Council Directive 79/409/EEC. 1979. *Birds Directive*. Available at: www.eur-lex.europa.eu/legal-content/EN/ALL/?uri=CELEX:31979L0409 [Accessed 1 September 2016].

Council Directive 92/43/EEC. 1992. *The Habitats Directive*. Available at: http://eur-lex.europa.eu/legal-content/EN/TXT/?uri=CELEX:31992L0043 [Accessed 1 September 2016].

Crane, J. 2013. *The Environment in American History: Nature and the Formation of the United States*. New York and London: Routledge.

Crang, M. and Tolia-Kelly, D. 2010. Nation, Race, and Affect: Senses and Sensibilities at National Heritage Sites. *Environ Plan A*, 42(10): 2315–2331.

Crawford, K.R. and Black, R. 2012. Visitor Understanding of the Geodiversity and the Geoconservation Value of the Giant's Causeway World Heritage Site, Northern Ireland. *Geoheritage*, 4(1): 115–126.

Cronon, W. 1997. *Uncommon Ground: Rethinking the Human Place in Nature*. New York: W.W. Norton.

Cuvier, G. 1813. *Essay on the Theory of the Earth* [Trans. R. Kerr]. Edinburgh: W. Blackwood.

Czerkas, S.M. and Olson, E.C. 1986. *Dinosaurs past and present*. Seattle and London: Natural History Museum of Los Angeles County in association with University of Washington Press.

Dant, T. 2012. *Television and the Moral Imaginary: Society through the Small Screen*. Basingstoke: Palgrave Macmillan.

Dante's Peak. 1997. R. Donaldson (Dir.). Universal Pictures [Film].

Darly, S. 2010. Agriculture et patrimoine identitaire des parcs naturels régionaux en Île-de-France: des situations contrastées. *Pour*, 205–206, 103–110.

Dartmoor National Park. 2016. *Natural Environment*. Available at: www.dartmoor.gov.uk/lookingafter/laf-naturalenv [Accessed 15 August 2016].

Darwin, C. 1859. *On the origin of species by means of natural selection, or the preservation of favoured races in the struggle for life*. London: John Murray.

Darwin, C. 1871. *The Descent of Man, and Selection in Relation to Sex*. London: John Murray.

Davies, G. 2000a. Science, Observation and Entertainment: Competing Visions of Postwar British Natural History Television, 1946–1967. *Cultural Geography*, 7(4): 432–460.

Davies, G. 2000b. Narrating the Natural History Unit: institutional orderings and spatial strategies. *Geoforum*, 31(4): 539–551.

Davis, P. 1996. *Museums and the Natural Environment: The Role of Natural History Museums in Biological Conservation*. Leicester: Leicester University Press.

Davis, P. 1999. *Ecomuseums: A Sense of Place*. London and New York: Leicester University Press/Continuum.

Dawdy, S.L. 2009. Millennial archaeology: locating the discipline in the age of insecurity. *Archaeological Dialogues*, 16(2): 131–142.

Debord, G. 1967. *La Société du spectacle*. Paris: Buchet-Chastel.

DEFRA. 2016. *National parks: 8-point plan for England (2016 to 2020)*. Available at: www.gov.uk/government/publications/national-parks-8-point-plan-for-england-2016-to-2020/title [Accessed 2 August 2016].

Deleuze, G. 1986. *Cinema 1: The Movement Image* [Trans. H. Tomlinson and B. Habberjam]. Minneapolis, MN: University of Minnesota Press.

Deleuze, G. 1988. *Bergsonism* [Trans. H. Tomlinson and B. Habberjam]. New York: Zone Books.

Deleuze, G. 1989. *Cinema 2: The Time Image* [Trans. H. Tomlinson and R. Galeta]. Minneapolis, MN: University of Minnesota Press.

Deleuze, G. 1994. Difference and Repetition [Trans. by P. Patton]. New York: Columbia University Press.

de Lumley, H. 1998. *L'Homme premier : préhistoire, évolution, culture*. Paris: Odile Jacob.

DeMarco, L. 2016. *Cleveland Museum of Natural History's iconic Steggie sculpture to be removed for repairs. March 15*. Available at: www.cleveland.com/entertainment/index.ssf/2016/03/cleveland_museum_of_natural_hi_2.html [Accessed 12 August 2016].

DeSalle, R. 1999. *Epidemic! The World of Infectious Disease*. New York: New Press.

Descartes, R. 1998. *Descartes: The World and Other Writings*. Edited by S. Gaukroger. Cambridge, England: Cambridge University Press.

Descola, P. 1994. *In the society of nature: a native ecology in Amazonia*. Cambridge, England: Cambridge University Press.

Descola, P. 2013. *Beyond Nature and Culture*. Chicago, IL: University of Chicago Press

Dickerson, M. 1914. New African Hall Planned by Carl E. Akeley. *Natural History: The American Museum Journal*, XIV: 175–189

Diehl, J. 2011. *Mammut-Prozess: Was Mladic in Den Haag erwartet*. Available at: www.spiegel.de/politik/ausland/mammut-prozess-was-mladic-in-den-haag-erwartet-a-765810.html [Accessed 1 July 2016].

Dinosaur Adventure. 2016. *Home*. Available at: www.dinosauradventure.co.uk [Accessed 2 June 2016].

Dinosaur Park. 2016. *Home*. Available at: www.thedinopark.com/ [Accessed 19 May 2016].

Dinosaur World. 2016a. *Home*. Available at: www.dinosaurworld.com [Accessed 20 May 2016].

Dinosaur World. 2016b. *Mammoths: Giants of the Ice Age*. Available at: www.dinosaurworld. com/florida/mammoths-giants-of-the-ice-age [Accessed 20 May 2016].

Dinosaurier Park. 2016a. *Über uns*. Available at: www.dinopark.de/Dinopark [Accessed 11 August 2016].

Dinosaurier Park. 2016b. *Pädagogik*. Available at: www.dinopark.de/Paedagogik/ [Accessed 11 August 2016].

Dinosaurs Alive! 2016a. *Meet the Dinos!* Available at: www.visitkingsisland.com/ dinosaursalive/meet-the-dinos [Accessed 15 May 2016].

Dinosaurs Alive! 2016b. *Scenes based on real fossil evidence*. Available at: www. kingsdominion.com/rides/Dinosaurs-Alive/Scenes-Based-on-Real-Fossil-Evidence [Accessed 15 May 2016].

Dinosaurs Unearthed. 2016. Homepage. Available at: http://www.dinosaursunearthed.com/ [Accessed 16 September 2016].

DinoZatorland. 2016. Available at: http://zatorland.pl/Zatorland-5.html [Accessed 20 August 2016].

Discovery. 2016. Home. Available at: www.discovery.com/ [Accessed 29 July 2016].

Dodds, F., Strauss, M. and Strong, M.F. 2012. *Only One Earth: The Long Road Via Rio to Sustainable Development*. London and New York: Routledge.

Doherty, B. and Doyle, T. (Eds.). 2008. *Beyond Borders: Environmental Movements and Transnational Politics*. London and New York: Routledge.

Dorfman, E. 2011. *Intangible Natural Heritage: New Perspectives on Natural Objects*. London and New York: Routledge.

Doughty, P. 2008. How things began: the origins of geological conservation. In C.V. Burek and C. D. Prosser (Eds.), *The History of Geoconservation* (7–16). London: The Geological Society.

Duarte, G.A. 2014. *Fractal Narrative. About the Relationship Between Geometries and Technology and Its Impact on Narrative Spaces*. Blelefeld: transcript Verlag.

Durst, S.B. and Barnum, P.T. 1849. *Sights and wonders in New York: including a description of the mysteries, miracles, marvels, phenomena, curiosities, and nondescripts, contained in that great congress of wonders, Barnum's Museum*. New York: J.S. Redfield.

Earthquake, 1974. M. Robson (Dir.). Universal Pictures [Film].

Ebeling, K.S. 2002. *Die Fortpflanzung der Geschlechterverhältnisse. Das metaphorische Feld der Parthenogenese in der Evolutionsbiologie*. Mössingen-Thalheim, Talheim: Verlag.

Eco, U. 1986. *Faith in Fakes*. London: Secker and Warburg.

Edwards, R. and Stewart, J. 1980. *Preserving indigenous cultures: a new role for museums. Papers from a regional seminar, Adelaide Festival Centre, 10–15 September 1978*. Canberra: Australian Government Publishing Service.

Eesti Loodusmuuseumi. 2016. *Home*. Available at: www.loodusmuuseum.ee/ [Accessed 7 July 2016].

EGN. 2000. *The EGN Charter*. Available at: www.europeangeoparks.org/?page_id=357 [Accessed 27 September 2016].

Eliot, C.W. 1870. *Address of Charles W. Eliot. First Annual Report of the American Museum of Natural History*. New York: American Museum of Natural History.

Elliot, N.L. 2010. Signs of Anthropomorphism: The Case of Natural History Television Documentaries. *Social Semiotics*, 11(3): 289–305.

Emerson, R.W. 1836. *Nature*. Boston: James Munroe and Company.

Engels, F. 1894. *Anti-Dühring; Herr Eugen Dühring's revolution in science*. New York: International Publishers.

Engels, F. 1940. *Dialectics of Nature* [Trans. C.P. Dutt]. New York: International Publisher.

Engström, K. and A.G. Johnels (Eds.). 1973. *Natural History Museums and the Community: Symposium Held in October 1969 at the Swedish Museum of Natural History (Naturhistoriska riksmuseet) in Stockholm*. Oslo: Universitetsforlaget.

Environment Act. 1995. *Environment Act*. Available at: www.legislation.gov.uk/ukpga/1995/25/coPlease insert:ntents [Accessed 7 September 2016].

Epstein, C. 2010. *Model Nazi: Arthur Greiser and the Occupation of Western Poland*. Oxford: Oxford University Press.

Eritja, M.C., Casado, L.C., Moreno, J.E.N., Solé, A.P. and Castejón, I.P. 2011. *Environmental Law in Spain*. Alphen aan de Rijn: Kluwer.

Espacios Naturales. 2004. La 2 Productions [Documentary].

Ettel, A., Seibel, K. and Zschäpitz, H. 2016. *Wer braucht heutzutage noch eine Bank?* Available at: www.welt.de/wirtschaft/article158496200/Wer-braucht-heutzutage-noch-eine-Bank.html [Accessed 3 October 2016].

European Landscape Convention. 2000. European Landscape Convention. Available at: https://rm.coe.int/CoERMPublicCommonSearchServices/DisplayDCTMContent?documentId=0900001680080621 [Accessed 2 August 2016].

Evans, D. 1992. *A History of Nature Conservation in Britain*. London and New York: Routledge.

Evermann, B.W. 1918. Modern Natural History Museums and their Relation to Public Education. *The Scientific Monthly,* 6(January): 5–36.

Fairclough, N. 1993. *Discourse and social change*. Cambridge: Polity Press.

Fairclough, N. 2001. *Language and Power*. London: Longman.

Feyerabend, P. 1975. *Against Method*. London: New Left Books.

Field Museum of Natural History. 1928. *Field Museum and the child. An outline of the work carried on by the Field museum of natural history among school children of Chicago through the N.W. Harris public school extension and the James Nelson and Anna Louise Raymond public school and children's lectures*. Chicago, IL: Field Museum of Natural History.

Field Museum of Natural History. 1976. March at the Field Museum. *Bulletin, January 1976,* 47(1): 18–19.

Field Station: Dinosaurs. 2016. *Home*. Available at: www.fieldstationdinosaurs.com [Accessed 29 March 2016].

Findsen, O. 1980. Art and Science Make a Museum. *Cincinnati Enquirer*, April 20.

Fleming, A. 2000. St Kilda: Family, Community, and the Wider World. *Journal of Anthropological Archaeology,* 19, 348–368.

Flower, W.H. 1898. *Essays on museums and other subjects connected with natural history*. London: Macmillan and Co.

Flying Elephants. 1928. F. Butler (Dir.). Pathé Exchange [Film].

Foucault, M. 1969. *Archaeology of Knowledge* [Trans. S. Smith]. New York: Pantheon.

Foucault, M. 1979. *Discipline and punish: The birth of the prison* [Trans. S. Smith]. New York: Vintage Books.

Foucault, M. 1991. Governmentality. In G. Burchell, C. Gordon and P. Miller (Eds.), *The Foucault Effect: Studies in Governmentality* (87–104). Hemel Hempstead: Harvester Wheatsheaf.

Fox, C. 2009. Dinosaur statues meticulously constructed. *Atlanta Journal-Constitution*, March 15.

Fox, D.M. and Karp, D.R. 1988. Images of Plague: Infectious Disease in the Visual Arts. In E. Fee and D.M. Fox (Eds.), *AIDS: The Burdens of History* (172–189). Berkeley, CA: University of California Press.

France 2. 2016. *Grandeurs Nature*. Available at: www.france2.fr/emissions/grandeurs-nature/une-excursion-hebdomadaire-dans-la-nature_309259 [Accessed 13 July 2016].

Freedman, S.H. 1984. Rift over Fossils from South Africa. *New York Times*, May 30.

Freud, S. 1913. *Totem and Taboo. Standard Edition XIII*. London: Hogarth Press and the Institute of Psycho-Analysis.

Frosh, P. 2006. Telling presences: Witnessing, mass media, and the imagined lives of strangers. *Critical Studies in Media Communication*, 23(4): 265–84.

Frosh, P. 2011. Telling Presences: Witnessing, Mass Media and the Imagined Lives of Strangers. In P. Frosh and A. Pinchevski (Eds.), *Media Witnessing: Testimony in the Age of Mass Communication* (49–72). Basingstoke: Palgrave Macmillan.

Frosh, P and Pinchevski, A. 2008. Introduction: Why Media Witnessing? Why Now? In P. Frosh and A. Pinchevski (Eds.), *Media Witnessing: Testimony in the Age of Mass Communication* (1–22). Basingstoke: Palgrave.

Fussell, P. 1975. *The Great War and Modern Memory*. Oxford: Clarendon Press.

Futter, V. 1997. Toward a Natural History Museum for the 21st Century. *Museum News*, 76(6): 41.

Gadamer, H.G. 1975. *Truth and Method*. New York: Crossroad.

Galápagos. 2006. BBC Productions [Documentary].

Galton, F. 1869. *Hereditary Genius: an inquiry into its laws and consequences*. London: Macmillan and Co.

Garçon, F. 2005. Le documentaire historique au péril du docufiction. *Vingtième Siècle. Revue d'histoire*, 4(88): 95–108.

Gauthier, G. 2011. *Le documentaire: un autre cinéma*. Paris: Armand Colin.

Gea Norvegica Geopark. 2016. *Geologiske hovedtrekk*. Available: www.geoparken.no/geologiske-hovedtrekk [Accessed 21 August 2016].

Gell, A. 1998. *Art and Agency: An Anthropological Theory*. Oxford: Clarendon Press.

Géopark des Bauges. 2016. *L'homme et la Terre*. Available at: www.parcdesbauges.com/fr/decouvrir-le-massif-des-bauges/s-emerveiller/l-homme-et-la-terre.html#.WAokougrLIU [Accessed 3 August 2016].

George S. Eccles Dinosaur Park. 2016. *Home*. Available at: www.dinosaurpark.org [Accessed 3 September 2016].

Gesellschaft Naturforschender Freunde zu Berlin. 1775. *Beschäftigungen, Vol.1*. Berlin: Joachim Pauli.

Giant's Causeway Official Guide. 2016. *Homepage*. Available at: www.giantscauseway officialguide.com/ [Accessed 1 August 2016].

Giddens, A. 1990. *The Consequences of Modernity*. Cambridge, England: Polity.

Giddens, A. 1991. *Modernity and Self-Identity: Self and Society in the Late Modern Age*. Stanford, CA: Stanford University Press.

Gillet, A. 2009. One Stone After the Other. Geopoetical Considerations on Stony Ground. In M. Smith, J. Davidson, L. Cameron and L. Bondi (Eds.), *Emotion, Place and Culture* (283–297). London, Ashgate..

Gissibl, B., Höhler, S. and Kupper, P. 2012. *Civilizing Nature: National Parks in Global Historical Perspective*. New York and Oxford: Berghahn.

Goatly, A. 2007. *Washing the Brain: Metaphor and Hidden Ideology*. Amsterdam: John Benjamins.

Goffman, E. 1972. *Relations in Public: Microstudies of the Public Order*. New York: Harper & Row.

Goggin, P.N. (Ed.). 2012. *Environmental Rhetoric and Ecologies of Place*. London and New York: Routledge.

Golding, V. 2009. *Learning at the Museum Frontiers: Identity, Race and Power*. Aldershot: Ashgate.

Gordon, J.E. 2012. Rediscovering a Sense of Wonder: Geoheritage, Geotourism and Cultural Landscape Experiences. *Geoheritage*, 4(65): 65–77.

Gordon, J.E. and Baker, M. 2016. Appreciating geology and the physical landscape in Scotland: from tourism of awe to experiential re-engagement. *Geological Society Special Publication*, 417: 25–40.

Gosling, D.C. 1980. Man's Place in Evolution: a new exhibition at the Natural History Museum. *Museums Journal*, 80(2): 66–9.

Gottlieb, R.S. (Ed.). 1997. *The Ecological Community*. London and New York: Routledge.

Goudie, A. 1981. *The human impact on the natural environment: past, present, and future*. Oxford: Blackwell.

Gould, S.J. 2007. *Punctuated Equilibrium*. Cambridge, MA: Belknap Press of Harvard University Press.

Gould, S.J. and Eldredge, N. 1977. Punctuated equilibria: the tempo and mode of evolution reconsidered. *Paleobiology*, 3(2): 115–151.

Graham, B. 2003. Interpreting the Rural in Northern Ireland: Place, Heritage and History. In J. Greer and M. Murray (Eds.), *Rural Planning and Development in Northern Ireland* (261–282). Dublin: Institute of Public Administration.

Grandeurs nature. 2009. France 2 Productions [Documentary].

Grew, N. 1681. *Musaeum regalis societatis*. London: s.n.

Gross, A. and Vallely, A. (Eds.). 2012. *Animals and the Human Imagination: A Companion to Animal Studies*. New York: Columbia University Press.

Groves, A. 2016. *Giant's Causeway*. Stroud: The History Press.

Guattari, F. 1989. The Three Ecologies. *New Formations*, 8: 131–147.

Guattari, F. 2000. *The Three Ecologies*. London: Athlone.

Guattari, F. 2008. *The Three Ecologies*. London: Continuum.

Guiomar, X. 2010. Éveiller le regard du marcheur sur l'agriculture: les sentiers d'interprétation agricole en Île-de-France. *Pour*, 205–206: 299–309.

Günther, A. 1885. *Guide to the galleries of Mammalia (mammalian, osteological, cetacean) in the Department of Zoology of the British Museum (Natural History)*. London: s.n.

Haldane, J.B.S. 1941. *Science in Peace and War*. London: Lawrence & Wishart.

Hallam, A. 1983. *Great Geological Controversies*. Oxford: Oxford University Press.

Hansen, A. 1991. The media and the social construction of the environment. *Media Culture Society*, 13(4): 443–458.

Haraway, D. 1984. Teddy Bear Patriarchy: Taxidermy in the Garden of Eden, New York City, 1908–1936. *Social Text*, 11: 20–64.

Haraway, D. 1989. *Primate Visions: Gender, Race and Nature in the World of Modern Science*. London and New York: Verso.

Haraway, D. 1991. *Simians, Cyborgs, and Women: The Reinvention of Nature*. London and New York: Routledge.

Haraway, D.J. 1997. *Modest_Witness@Second_Millennium.Femaleman_Meets_Oncomouse: Feminism and Technoscience*. New York: Routledge.

Haraway, D. 2003. *The Companion Species Manifesto: Dogs, People and Significant Otherness*. Chicago, IL: Prickly Paradigm Press.

Haraway, D. 2008. *When Species Meet*. Minneapolis, MN: University of Minnesota Press.

Haraway, D. 2015. Anthropocene, Capitalocene, Plantationocene, Chthulucene: Making Kin. *Environmental Humanities*, 6: 159–165.

Haraway, D. 2016. *Staying with the Trouble: Making Kin in the Chthulucene*. Durham, NC: Duke University Press.

Harré, R. 1993. *Laws of Nature*. London: Duckworth.

Harrison, R. 2009. *Understanding the Politics of Heritage*. Manchester: Manchester University Press.

Harrison, R. 2013. *Heritage: Critical Approaches*. Abingdon: Routledge.

Harrison, R. 2015. Beyond "Natural" and "Cultural" Heritage: Toward an Ontological Politics of Heritage in the Age of Anthropocene. *Heritage & Society*, 8(1): 24–42.

Harrison, R., Fairclough, G., Schofield, J. and Jameson, J.H. 2008. Heritage, memory and modernity: An introduction. In G.J. Fairclough, R. Harrison, J. Schofield and J.H. Jameson (Eds.), *The Heritage Reader* (1–12). London and New York: Routledge.

Hartley, J. 1999. *Uses of television*. London and New York: Routledge.

Harvey, D. 1996. *Justice, Nature, and the Geography of Difference*. Cambridge, MA: Blackwell.

Harvey, D.C. and Perry, J. (Eds.). 2015. *The Future of Heritage as Climates Change: Loss, Adaptation and Creativity*. Abingdon and New York: Routledge.

Hatley, J. 2000. *Suffering Witness: The Quandary of Responsibility after the Irreparable*. Albany, NY: SUNY Press.

Hawley, K. 2002. *How things persist*. Oxford: Clarendon Press.

Hebenstreit, J.E. and Christ, J.F. 1743. *Museum Richterianum continens fossilia animalia vegetabilia mar*. Liepzig: Casparus Fritsch.

Heidegger, M. 1962. *Being and Time* [Trans. by J. MacQuarrie and E. Robinson]. New York: Harper.

Heidegger, M. 1995. *The Fundamental Concepts of Metaphysics: World, Finitude, Solitude*. Bloomington, IN: Indiana University Press.

Hein, H. 2000. *The Museum in Transition: A Philosophical Perspective*. Washington, DC: Smithsonian Institution Press.

Heise, U.K. 2008. *Sense of Place and Sense of Planet: The Environmental Imagination of the Global*. Oxford: Oxford University Press.

Henare, A., Holbraad, A. and Wastell, S. 2007. *Thinking Through Things: Theorising Artefacts Ethnographically*. London: Routledge.

Hewison, R. 1987. *The heritage industry: Britain in a climate of decline*. London: Methuen.

Hirschauer, S. 2005. On Doing Being a Stranger: The Practical Constitution of Civil Inattention. *Journal for the Theory of Social Behaviour*, 35(1): 41–67.

His Prehistoric Past. 1914. C. Chaplin (Dir.). Keystone Studios [Film].

Histoires naturelles. 1981–2009. TF1 Productions [Documentary].

Hitchcock, E. 1865. *Supplement to the Ichnology of New England: A Report to the Government of Massachusetts*. Boston, MA: Wright and Potter.

Holland, J.A. 1917. A Canadian's Stirring Battle Picture. *The New York Times Current History: The European War*, 10: 690–693.

Homo Sapiens. 2005. France 3 Productions [Documentary].

Howard, P. and Luginbühl, Y. (Eds.). 2015. *Landscape and Sustainable Development: The French Perspective*. London and New York: Routledge.

Howard, P. and Papayannis, T. (Eds.). 2007. *Natural Heritage: At the Interface of Nature and Culture*. London and New York: Routledge.

Howard, P. and Pinder, D. 2003. Cultural heritage and sustainability in the coastal zone: experiences in south west England. *Journal of Cultural Heritage*, 4(1): 57–68.

Huizinga, J. 1949. *Homo Ludens: a study of the play-element in culture*. London: Routledge and Kegan Paul.

Hunter, J.R. and Smith, Z. 2005. *Protecting Our Environment: Lessons from the European Union*. Albany, NY: SUNY Press

Hupke, K-D. 2015. *Naturschutz: Ein kritischer Ansatz*. Dordrecht: Springer.

Husserl, E. 1991. *On the Phenomenology of the Consciousness of Internal Time (1893–1917)* [Trans. J. Bamett Brough]. Edited by Rudolf Bemet. Dordrecht: Kluwer.

Hussey, S. and Thompson, P. 2000. *The Roots of Environmental Consciousness: Popular Tradition and Personal Experience*. London: Routledge.

Hutchinson, H.N. 1897. *Extinct Monsters*. London: Chapman and Hall.

Hutt, S., Blanco, C.M. and Varmer, O. 1999. *Heritage Resources Law: Protecting the Archeological and Cultural Environment*. New York: John Wiley & Sons, Inc.

Hutterer, R. 2014. Habitat dioramas as historical documents: A case study. In S. Tunnicliffe and A. Scheersoi (Eds.), *Natural History Dioramas* (23–32). Berlin: Springer.

Hutton, J. 1788. *Theory of the Earth; or an investigation of the laws observable in the composition, dissolution, and restoration of land upon the Globe*. Edinburgh: Transactions of the Royal Society of Edinburgh.

Huxley, T.H. 1863. *Evidence as to Man's Place in Nature*. New York: D. Appleton.

Ice Age Giants. 2013. BBC Productions [Documentary].

ICOM. 2013. *ICOM Code of Ethics for Natural History Museums*. Paris: ICOM.

Imperato, F. 1599. *Dell'Historia Naturale*. Naples: s.n.

In Birdland. 1907. O.G. Pike (Photographer) [Film].

Ingleborough Cave. 2016. *About Ingleborough Cave*. Available at: www.ingleboroughcave.co.uk/about.htm [Accessed 2 July 2016].

Inglis, D., Bone, J. and Wilkie, R. 2005. *Nature: Thinking the natural, Volume 1*. London and New York: Routledge.

Ingold, T. 2000. *The Perception of the Environment*. London and New York: Routledge.

Ingold, T. and Vergunst, J.L. 2008. Introduction. In T. Ingold and J.L. Vergunst (Eds.), *Ways of Walking: Ethnography and Practice on Foot* (1–19). Burlington, VT: Ashgate.

Inman, P. 2016. Philip Hammond could face £84bn black hole following Brexit vote. *The Guardian*, 26 October.

Institut Royal des Sciences Naturelles de Belgique. 2016. *Galerie de l'Homme*. Available at: www.naturalsciences.be/fr/museum/exhibitions-view/771/2762/377 [Accessed 7 July 2016].

Jackson, J.B. 1994. *A Sense of Place, a Sense of Time*. New Haven, CT: Yale University Press.

Jardine, N., Secord, J.A. and Spary, E.C. 1996. *Cultures of Natural History*. Cambridge, England: Cambridge University Press.

Jefferies, M. 2010. BBC natural history versus science paradigms. *Science as Culture*, 12(4): 527–545.

Jones, M. and Stenseke, M. 2011. *The European Landscape Convention: Challenges of Participation*. Dordrecht: Springer.

Joyce, E.B. 1994. Geological Heritage Committee. In B.J. Cooper and D.F. Branagan (Eds.), *Rock me hard. Rock me soft. A history of the Geological Society of Australia Inc* (30–36). Sydney: Geological Society of Australia Incorporated.

JuraPark. 2016. *Home.* Available at: www.juraparkbaltow.pl [Accessed 1 July 2016].

Jurassic Coast. 2016. *What is the Jurassic Coast.* Available at: www.jurassiccoast.org/about/what-is-the-jurassic-coast [Accessed 6 August 2016].

Jurassic Park. 1993. S. Spielberg (Dir.). Universal Pictures [Film].

Kant, I. 1998. *Critique of Pure Reason.* Cambridge, England: Cambridge University Press.

Kavanagh, G. 1994. *Museums and the First World War: a social history.* Leicester: Leicester University Press.

Keane, S. 2001. *Disaster Movies: The Cinema of Catastrophe.* London: Wallflower Press.

Keller, D.R. and Golley, F.B. (Eds.). 2000. *The Philosophy of Ecology: From Science to Synthesis.* Athens, GA: University of Georgia Press.

Kenosha Natural History Museum. 2015. *Home.* Available at: www.kenosha.org/wp-museum/exhibits-2/permanent-exhibits/ [Accessed 9 July 2016].

Kidd, J., Cairns, S., Drago, A., Ryall, A. and Stearn, M. (Eds.). 2014. *Challenging History in the Museum: International Perspectives.* Abingdon and New York: Routledge.

Killingsworth, M.J. and Palmer, J.S. 1992. *Ecospeak: Rhetoric and Environmental Politics in America.* Carbondale and Edwardsville, IL: Southern Illinois University Press.

King Kong. 1933. M.C. Cooper and E.B. Schoedsack (Dirs.). Radio Pictures [Film].

Kiss, A. 1999. Nature: the common heritage of mankind. *Naturopa,* 91: 10–12.

Kohn, E. 2013. *How Forests Think: Towards an Anthropology Beyond the Human.* Berkeley, CA: University of California Press.

Kohn, E. 2015. Anthropology of Ontologies. *Annual Review of Anthropology,* 44: 311–327.

Konstytucja, RP. 1997. *Dziennik Ustaw,* 78(483): 31.

Kovecses, Z. 2002. *Metaphor: A Practical Introduction.* Oxford: Oxford University Press.

Krishtalka, L. and Humphrey, P.S. 2000. Can Natural History Museums Capture the Future? *BioScience,* 50(7): 611–617.

Kuhn, P. 1962. The Structure of Scientific Revolutions. Chicago, IL: University of Chicago Press.

Kulturmiljölag. 1988. *950.* Available at: www.notisum.se/rnp/sls/lag/19880950.htm [Accessed 17 August 2016].

L'odyssée de l'espèce. 2003. France 3 Productions [Documentary].

La vraie vie des dinosaurs. 2012. Explora Productions [Documentary].

LaBRIGS. 2006. *James Hutton: a man ahead of his time.* Available at: www.edinburghgeolsoc.org/downloads/rigsleaflet_huttona4.pdf [Accessed 20 June 2016].

Laclau, E. 1990. *New Reflections on the Revolution of Our Time.* London: Verso.

Lakoff, G. and Johnson, M. 1980. *Metaphors We Live By.* Chicago, IL: The University of Chicago Press.

Lamarck, J-B. 1801. *Système des animaux sans vertèbres.* Paris: Chez l'auteur au Muséum d'Hist. Naturelle.

Lamarck, J-B. 1809. *Philosophie zoologique, ou exposition.* Paris: Chez l'auteur au Muséum d'Hist. Naturelle.

Lamb, R. 1996. *Promising the Earth.* London and New York: Routledge.

Lampert, R.J. 1986. The Development of the Aboriginal Gallery at the Australian Museum. *Bulletin of the Conference of Museum Anthropologists,* (18): 10–18.

Land of the Tiger. 1997. BBC Productions [Documentary].

Landes, J.B., Young, P. and Youngquist, P. 2012. *Gorgeous Beasts: Animal Bodies in Historical Perspective.* Philadelphia: Penn State Press.

Langston, W. 1960. The vertebrate fauna of the Selma Formation of Alabama. Part VI. The dinosaurs. *Fieldiana: Geology Memoirs,* 3(6): 315–361.

Larrère, R. 2009. Histoire(s) et mémoires des parcs nationaux. In R. Larrère, B. Lizet and M. Berlan-Darqué (Eds.), *Histoire des parcs nationaux : Comment prendre soin de la nature?* (21–41). Paris: Editions Quæ.

Latour, B. 1987. *Science in Action: How to Follow Scientists and Engineers through Society.* Cambridge, MA: Harvard University Press.

Latour, B. 1993. *We have never been modern.* Cambridge, MA: Harvard University Press.

Latour, B. 1998. To modernize or to ecologize? That's the question. In N. Castree and B. WillemsBraun (Eds.), *Remaking Reality: Nature at the Millennium* (221–242). London and New York: Routledge.

Latour, B. 2004a. *Politics of Nature: How to Bring the Sciences into Democracy.* Cambridge, MA: Harvard University Press.

Latour, B. 2004b. Why Has Critique Run out of Steam? From Matters of Fact to Matters of Concern. *Critical Inquiry,* 30(2): 225–248.

Latour, B. 2005. *Making Things Public: Atmospheres of Democracy.* Cambridge, MA: MIT Press.

Latour, B. 2014. Another way to compose the common world. *HAU: Journal of Ethnographic Theory,* 4(1): 301–307.

Laughlin, H.H. 1923. *The second International Exhibition of Eugenics held September 22 to October 22, 1921, in connection with the Second International Congress of Eugenics in the American Museum of Natural History, New York.* Baltimore, MD: Williams and Wilkins Company.

Le monde du silence. 1956. J. Cousteau and L. Malle (Dirs.). FSJYC Production [Film].

Le Monde sans soleil. 1964. J. Cousteau (Dirs.). Columbia Pictures [Film].

Le Sacre de l'homme. 2007. France 2 Productions [Documentary].

Lee, R. 1820. *Taxidermy: or, The art of collecting, preparing, and mounting objects of natural history. For the use of museums and travellers.* London: Longman, Hurst, Rees, Orme, and Brown.

Lefevre, G.W. 1843. *The Life of a Travelling Physician: From His First Introduction to Practice, Vol. III.* London: Longman, Brown, Green and Longmans.

Legge 6–12–1991 n. 394. 1991. *Legge quadro sulle aree protette.* Available at: http://www.gazzettaufficiale.it/eli/id/1991/12/13/091G0441/sg [Accessed 4 August 2016].

Lekan, T.M. 2004. *Imagining the Nation in Nature: Landscape Preservation and German Identity, 1885–1945.* Cambridge, MA: Harvard University Press.

Lester, L. 2010. *Media and Environment: Conflict, Politics and the News.* Cambridge, England: Polity Press.

Lester, L. and Cottle, S. 2009. Visualizing Climate Change: Television News and Ecological Citizenship. *International Journal of Communication,* 3: 920–936.

Levine, L.D. (Ed.). 1975. *Man in nature: historical perspectives on man in his environment.* Toronto: Royal Ontario Museum.

Lewis, D.K. 1986. *On the Plurality of Worlds.* Oxford: Blackwell.

Ley 4/1989. 1989. *Ley de Conservación de los Espacios Naturales y de la Flora y Fauna Silvestres.* Available at: https://www.boe.es/buscar/doc.php?id=BOE-A-1989–6881 [Accessed 7 August 2016].

Ley 42/2007. 2007. del Patrimonio Natural y de la Biodiversidad. Available at: https://www.boe.es/buscar/act.php?id=BOE-A-2007-21490 [Accessed 13 August 2016].

Life on Earth. 1979. BBC Productions [Documentary].

Life in the Air. 2016. BBC Productions [Documentary].

Life in the Freezer. 1993. BBC Productions [Documentary].

Linneaus, C. 1735. *Systema naturae; sive, Regna tria naturae: systematice proposita per classes, ordines, genera, & species.* Leiden: Joannis Wilhelmi de Groot.

Linneaus, C. 1764. *Museum Ludovica Ulrica Reginae Suecorum.* Holmiae: Laur. Salvii.

Living Britain. 1999. BBC Productions [Documentary].

Locke, J. 1836. *An Essay Concerning Human Understanding.* London: T. Tegg and Son.

Loi n°60–708. 1960. *Loi du 22 juillet 1960 relative à la création de parcs nationaux, 1960.* Available at: https://www.legifrance.gouv.fr/affichTexte.do?cidTexte=JORFT EXT000000512209 [Accessed 8 July 2016].

Loi n° 2006–436. 2016. *Loi du 14 avril 2006 relative aux parcs nationaux, aux parcs naturels marins et aux parcs naturels régionaux.* Available at: https://www.legifrance. gouv.fr/eli/loi/2006/4/14/DEVX0500070L/jo/texte [Accessed 8 July 2016].

Looby, C. 1987. The constitution of nature: taxonomy as politics in Jefferson, Peale, and Bartram. *Early American Literature,* 22(3): 259–264.

Loohauis, J. 2007. Public museum makes a mammoth catch. *Milwaukee Wisconsin Journal Sentinel,* July 31.

Lord, B. 2006. Foucault's museum: Difference, representation, and genealogy. *Museum and Society,* 4(1): 11–14.

Lovelock, J.E. and Giffin, C.E. 1969. Planetary Atmospheres: Compositional and other changes associated with the presence of Life. *Advances in the Astronautical Sciences,* 25: 179–193.

Lovelock, J.E. and Margulis, L. 1974. Atmospheric homeostasis by and for the biosphere: the Gaia hypothesis. *Tellus. Series A,* 26(1–2): 2–10.

Lowenthal, D. 1985. *The Past is a Foreign Country.* Cambridge, England: Cambridge University Press.

Lowenthal, D. and Olwig, K. (Eds.). 2006. *The Nature of Cultural Heritage, and the Culture of Natural Heritage.* Abingdon and New York: Routledge.

Lucas, F.A. 1901. *Animals of the past: an account of some of the creatures of the ancient world.* New York: American Museum of Natural History.

Lucas, F.A. 1921. *The story of museum groups.* New York: American Museum Press.

Lyell, C. 1832. *Principles of geology, being an attempt to explain the former changes of the Earth's surface, by reference to causes now in operation.* London: John Murray.

Lyons, J. 2010. Fury at arrogant Tory Sir Nicholas Winterton's refusal to travel standard class on trains. *Daily Mirror,* 19 February.

Lyotard, J-F. 1984. *The Postmodern Condition: A Report on Knowledge* [Trans. G. Bennington and B. Massumi]. Minneapolis, MN: University of Minnesota Press.

Lyotard, J-F. 1988. *Le Différend.* Minneapolis, MN: The University of Minnesota Press.

Macdonald, S. 1998. Exhibitions of Power and Powers of Exhibitions: An Introduction to the Politics of Display. In S. Macdonald (Ed.), *The Politics of Display: Museums, Science, Culture* (1–24). New York and London: Routledge.

Macdonald, S. 2002. *Behind the Scenes of the Science Museum.* Oxford: Berg.

Machin, R. 2008. Gender representation in the natural history galleries at the Manchester Museum. *Museum and Society,* 6(1):54–67.

Mackensen, L. 1943. *Sagen der Deutschen im Wartheland.* Posen: Hirt-Reger und v.Schroedel-Siemau.

MacLeod, S. 2005. *Reshaping Museum Space: Architecture, Design, Exhibitions.* London and New York: Routledge.

MacMahon, D. 2013. Entrance Icons: Visual Meaning-Making in Museum Entrance Galleries. *Exhibitionist,* Spring 2013: 36–41.

Macnaghten, P. and Urry, J. 1998. *Contested Natures*. London: Sage.

Man's Genesis. 1912. D.W. Griffith (Dir.). Biograph Company [Film].

Man's Place in Evolution. 1980. *Exhibition Catalogue*. London: British Museum (Natural History) / Cambridge, England: Cambridge University Press.

Mance, H. and Blitz, J. 2016. Whitehall boosted by £412m extra funding for Brexit talks. *Financial Times*, November 23.

Manesse, D.J. 1787. *Traité sur la manière d'empailler et de conserver les animaux*. Paris: chez Guillot.

Margalit, A. 2002. *The Ethics of Memory*. Cambridge, MA: Harvard University Press.

Marsh, C. 2010. *Meet the Simi Valley Mastodon*. Available at: www.simivalleyacorn.com/news/2010-07-16/Front_Page/Meet_the_Simi_Valley_Mastodon.html [Accessed 10 July 2016].

Marsh, T. 2001. *Lake District (Official National Park Guide)*. Bath: Crimson Publishing.

Matthew, W.D. 1916. Lessons from Dinosaurs. *New York Times*, April 9.

May, J.R. and Daly, E. 2014. *Global Environmental Constitutionalism*. Cambridge, England: Cambridge University Press.

McLuhan, M. 1964. *Understanding Media. The extensions of man*. London and New York: Routledge.

McGrath, A. and Jebb, M. (Eds.). 2015. *Long history, deep time: deepening histories of place*. Canberra: Australia National University Press.

McGregor, J.C. 1943. The role of the museum in war. *The Living Museum*, 5(3): 23.

McKeever, P.J. and Zouros, N. 2005. Geoparks: Celebrating earth heritage, sustaining local communities. *Episodes*, December: 274–278.

McManus, R. 2009. Heritage and Tourism in Ireland – an unholy alliance? *Irish Geography*, 30(2): 90–98.

Meister, M and Japp, P.M. 2002. *Enviropop: Studies in Environmental Rhetoric and Popular Culture*. Westport, CT: Greenwood.Melosi, M.V. and Scarpino, P. (Eds.). 2004. *Public History and the Environment*. Malabar, FL: Krieger Publishing Company.

Mels, T. 2002. Nature, Home, and Scenery: The Official Spatialities of Swedish National Parks. *Environment and Planning D*, 20(2): 135–154.

Meringolo, D.D. 2012. *Museums, Monuments, and National Parks: Toward a New Genealogy of Public History*. Amherst, MA: University of Massachusetts Press.

Merrick, J. 2008. Now MPs hound Speaker because of the way he is handling expenses row. *Daily Mail*, 12 February.

Metzler, S. 2008. *Theatres of Nature: Dioramas at the Field Museum*. Chicago, IL: University of Chicago Press.

Michigan State University Museum. 2016. *Hall of Evolution*. Available at: http://museum.msu.edu/?q=node/160 [Accessed 19 August 2016].

Miles, E. 2013. Involving local communities and volunteers in geoconservation across Herefordshire and Worcestershire, UK – the Community Earth Heritage Champions Project. *Proceedings of the Geologists' Association*, 124(4): 691–698.

Miles Head, R.S. and Tout, A.F. 1978. Human Biology and the New Exhibition Scheme in the British Museum (Natural History). *Curator: The Museum Journal*, 21(1): 36–50.

Sveriges Riksdag 1998. Miljöbalk. Available at: www.notisum.se/rnp/sls/lag/19980808.htm [Accessed 3 August 2016].

Mill, J.S. 1874. *Nature, the Utility of Religion, and Theism*. London: Longmans, Green, Reader, and Dyer.

Miller, M.A. 2011. *A Natural History of Revolution: Violence and Nature in the French Revolutionary Imagination, 1789–1794*. Ithaca, NY: Cornell University Press.

Miller, S.L. 2000. Sue Makes Gargantuan Debut: Field Museum Unveils Tyrannosaurus Rex Fossil. *Chicago Tribune,* May 17.

Minkowski, E. 1970. *Lived Time* [Trans. N. Metzel]. Evanston: Northwestern University Press.

Mitchell, W.J.T. 1998. *The Last Dinosaur Book: The Life and Times of a Cultural Icon.* Chicago, IL: Chicago University Press.

Mitman, G. 1993. Cinematic Nature: Hollywood Technology, Popular Culture, and the American Museum of Natural History. *Isis,* 84: 637–661.

Mitman, G. 1999. *Reel Nature: America's Romance with Wildlife on Film.* Cambridge, MA: Harvard University Press.

MNHN. 2016. *Home.* Available at: www.mnhn.fr/fr/visitez/lieux/grande-galerie-evolution. Accessed 15 November 2016. [Accessed 27 July 2016].

Mohen, J-P. 2004. *Le Nouveau Musée de l'Homme.* Paris: Odile Jacob.

Montaño, M. 1999a. *Periodismo ambiental en Canal Sur Televisión.* Available at: www.revistalatinacs.org/a1999iab/100A/montano.html [Accessed 12 August 2016].

Montaño, M. 1999b. El Canal 2 de Andalucía y la Información Ambiental: el Programa Espacio Protegido. *Ámbitos,* 2: 207–227.

Morton, T. 2007. *Ecology Without Nature: Rethinking Environmental Aesthetics.* Cambridge, MA: Harvard University Press.

Morton, T. 2010. *The Ecological Thought.* Cambridge, MA: Harvard University Press.

Morton, T. 2013. *Hyperobjects: Philosophy and Ecology after the End of the World.* Minneapolis, MN: University of Minnesota Press.

Morton, T. 2016. *Dark Ecology: For a Logic of Future Coexistence.* New York: Columbia University Press.

Mosley, S. 2006. Common Ground: Integrating Social and Environmental History. *Journal of Social History,* 39(3): 915–933.

MSNB/KBIN. 2016. *Missions.* Available at: www.naturalsciences.be/en/about-us/mission/missions [Accessed 21 August 2016].

Mullen, W. 2008. Field Museum unveils timely nature exhibit. *Chicago Tribune,* May 20.

Murphy, B. 2016. *Museums, Ethics and Cultural Heritage.* Abingdon and New York: Routledge.

Murphy, J.A. 1960. Department of Geology. In Denver Museum of Natural History, *Annual Report* (17–18). Denver: Denver Museum of Natural History.

Musée de l'Homme. 2016. *Home.* Available at: www.museedelhomme.fr/fr/visitez/espaces/qui-sommes-nous [Accessed 19 July 2016].

Musée-Parc des Dinosaures. 2016. *Home.* Available at: www.dinosaure.eu [Accessed 10 August 2016].

Museo de la Evolución Humana. 2016. *Exposición permanente: Las joyas de la humanidad han venido para quedarse.* Available at: www.museoevolucionhumana.com/es/exposicion-permanente [Accessed 16 August 2016].

Museo Nacional de Ciencias Naturales. 2016a. *Home.* Available at: www.mncn.csic.es/Menu/Elmuseo/Presentacinehistoria/seccion=1177&idioma=es_ES.do [Accessed 13 June 2016].

Museo Nacional de Ciencias Naturales. 2016b. *Minerales, Fósiles y Evolución Humana.* Available at: www.mncn.csic.es/Menu/Exposiciones/Permanentes_Minerales_Fosiles_y_Evolucion_Humana/seccion=1182&idioma=es_ES&id=2010062410380001&activo=11.do [Accessed 13 June 2016].

Museum für Naturkunde. 2016a. *Home.* Available at: www.naturkundemuseum.berlin/de/museum/ausstellungen/mineralien [Accessed 14 June 2016].

Museum für Naturkunde. 2016b. *Kosmos & Sonnensystem*. Available at: www.natur kundemuseum.berlin/de/museum/ausstellungen/kosmos-sonnensystem [Accessed 14 June 2016].

Muséum National d'Histoire Naturelle. 2016. *Missions*. Available at: www.mnhn.fr/fr/ propos-museum/missions [Accessed 19 June 2016].

Museum of Natural History, University of Michigan. 2016. *Planetarium*. Available at: https://lsa.umich.edu/ummnh/planetarium/about-the-planetarium.html [Accessed 22 August 2016].

Museums in Australia. 1975. *Report of the Committee of Inquiry on Museums and National Collections*. Canberra: Australian Government Publishing Service.

Myerson, G. and Rydin, Y. 1996. *The Language Of Environment: A New Rhetoric*. London and New York: Routledge.

Næss, A. 1989. *Ecology, community, and lifestyle*. Cambridge, England: Cambridge University Press.

National Geographic. 2016. *Home*. Available at: www.channel.nationalgeographic.com/ shows/ [Accessed 1 August 2016].

National Museum of Australia Act. 1980. *Section 3*. Canberra. Australian Government.

National Museum of Australia. 2001. *Building History: The National Museum of Australia*. Canberra: National Museum of Australia.

National Parks. 2016. *Home*. Available at: www.nationalparks.gov.uk/ [Accessed 1 August 2016].

National Parks and Access to the Countryside Act. 1949. *Chapter 97*. London: The Stationery Office.

National Parks Service Act. 1916. *An Act to establish a National Park Service, and for other purposes. 39 Stat. 535*. Washington, DC: Government Printing Office.

National Show Caves Centre. 2016. Homepage. Available at: www.showcaves.co.uk [Accessed 20 August 2016].

Nationalpark Eifel. 2016a. *Wandern an der Seite von Experten*. Available at: www. nationalpark-eifel.de/go/eifel/german/Gefuehrtes_Wandern/Gefuehrtes_Wandern.html [Accessed 22 March 2016].

Nationalpark Eifel. 2016b. *Geführte Wanderungen auf dem Schöpfungspfad*. Available at: www.nationalpark-eifel.de/go/eifel/german/Gefuehrtes_Wandern/Spirituelle_ Wanderungen.html [Accessed 22 March 2016].

Natura. 2000. *Home*. Available at: http://ec.europa.eu/environment/nature/natura2000/ sites_hab/index_en.htm [Accessed 11 August 2016].

Natural England. 2008. *Annual Report and Accounts 1 April 2007 to 31 March 2008*. Available at: www.gov.uk/government/uploads/system/uploads/attachment_data/file/ 248555/0727.pdf [Accessed 13 August 2016].

Natural England. 2013a. *Kingley Vale: National Nature Reserve – Nature Trail*. Worcester: Natural England.

Natural England. 2013b. *Walberswick: National Nature Reserve – Nature Trail*. Worcester: Natural England.

Natural History Museum. 2011. *Sexual Nature*. London: Natural History Museum.

Natural History Museum. 2016a. *Our Vision and Strategy*. Available at: www.nhm.ac.uk/ about-us/our-vision-strategy.html [Accessed 19 August 2016].

Natural History Museum. 2016b. *Human Evolution*. Available at: www.nhm.ac.uk/visit/ galleries-and-museum-map/human-evolution.html [Accessed 14 June 2016].

Natural History Museum. 2016c. *Earthquakes and Volcanoes*. Available at: www.nhm.ac.uk/visit/galleries-and-museum-map/volcanoes-and-earthquakes.html [Accessed 29 July 2016].

Natural History Museum of Crete. 2016. *Earthquake simulator*. Available at: www.nhmc.uoc.gr/el/exhibition-halls/earthquake-simulator [Accessed 13 July 2016].

Natural History Museum of Los Angeles County 2016a. *About*. Available at: www.nhm.org/site/about-our-museums/mission [Accessed 17 July 2016].

Natural History Museum of Los Angeles County 2016b. *Age of Mammals*. Available at: www.nhm.org/site/explore-exhibits/permanent-exhibits/age-of-mammals/exhibit-highlights [Accessed 16 May 2016].

Natural History Museum of Los Angeles County 2016c. *Mineral Hall*. Available at: www.nhm.org/site/sites/default/files/pdf/gallery_guides/NHM_Mineral_Hall_Gallery_Guide.pdf [Accessed 16 May 2016].

Natural History Museum of Utah. 2016. *About*. Available at: https://nhmu.utah.edu/museum/about/mission-values [Accessed 23 July 2016].

Naturhistorisches Museum Wien. 2016a. *Home*. Available at: www.nhm-wien.ac.at/museum [Accessed 22 June 2016].

Naturhistorisches Museum Wien. 2016b. *Anthropology*. Available at: www.nhm-wien.ac.at/ausstellung/dauerausstellung__schausammlung/hochparterre/saal_14-15_anthropologie [Accessed 22 June 2016].

Naturhistorisches Museum Wien. 2016c. *Minerals*. Available at: www.nhm-wien.ac.at/ausstellung/dauerausstellung__schausammlung/hochparterre/saal_1-5_mineralien_gesteine_meteoriten [Accessed 22 June 2016].

Naturhistorisches Museum Wien. 2016d. *Planetarium*. Available at: www.nhm-wien.ac.at/planetarium [Accessed 22 June 2016].

Naturhistorisk museum. 2006. *Mot naturens orden: En utstilling om homoseksualitet i dyreriket*. Oslo: Naturhistorisk museum.

Naturhistoriska riksmuseet. 2016. *Den mänskliga resan*. Available at: www.nrm.se/besokmuseet/utstallningar/denmanskligaresan.5713.html [Accessed 29 July 2016].

Naturvårdslag. 1964. *822*. Available at: www.notisum.se/rnp/sls/lag/19640822.htm [Accessed 30 June 2016].

Naturwissenschaftliche Museum. 2016. *Die Sammlung*. Available at: www.museen-aschaffenburg.de/Naturwissenschaftliches--Museum-/Sammlung/DE_index_1396.html [Accessed 30 June 2016].

NDR. 2016. *Expeditionen ins Tierreich*. Available at: www.ndr.de/fernsehen/sendungen/expeditionen_ins_tierreich/ [Accessed 17 August 2016].

NDR. 2016. *Wilde Heimat*. NDR Television [Documentary].

Nederlands Centrum voor Biodiversiteit Naturalis. 2016. *Homepage*. Available at: www.naturalis.nl/nl/ [Accessed 31 July 2016].

Nicholson, T.S. 1984. *Director's Message. American Museum of Natural History, 115th Annual Report*. New York: American Museum of Natural History.

Nora, P. (Ed.). 1984. *Les lieux de mémoire, Vol. 1*. Paris: Gallimard.

Norrell, M. Gaffney, E.S. and Dingus, L. 1995. *Discovering Dinosaurs in the American Museum of Natural History*. New York: Alfred A. Knopf.

Northumberland National Park. 2016. *Ice Age glaciation*. Available at: www.northumberlandnationalpark.org.uk/activities/ice-age-glaciation/ [Accessed 17 July 2016].

Noys, B. 2014. *Malign Velocities: Accelerationism and Capitalism*. Winchester: Zero Books.

NPS. 2015. *America's Geological Heritage: An Invitation to Leadership*. Available at: www.nature.nps.gov/geology/geoheritage/docs/GH_Publicaton_Final.pdf [Accessed 1 August 2016].

Nunès, E. 2015. L'éducation nationale est "un dinosaure décrépit" mais "généreux". *Le Monde*, 31 July.

Nyhart, L.K. 2009. *Modern Nature: The Rise of the Biological Perspective in Germany*. Chicago, IL: University of Chicago Press.

O'Halloran, D., Green, C., Harley, M., Stanley, M. and Knil, J. (Eds.). 1994. *Geological and Landscape Conservation. Proceedings of the Malvern International Conference 1993*. London: Geological Society.

Oelschlaeger, M. 1993. *The Idea of Wilderness: From Prehistory to the Age of Ecology*. New Haven, CT: Yale University Press.

Oglivie, B.W. 2006. *The Science of Describing: Natural History in Renaissance Europe*. Chicago, IL: Chicago University Press.

Olwig, K.R. 2010. 'Time Out of Mind–Mind Out of Time': custom versus tradition in environmental heritage research and interpretation. *International Journal of Heritage Studies*, 4(7): 339–354.

Osborn, H.F. 1910. *History, plan and scope of the American Museum of Natural History*. New York: Irving Press.

Osborn, H.F. 1916. *Men of the Old Stone Age: Their Environment, Life and Art*. New York: Charles Scribner's Sons.

Osborn, H.F. 1920. Nature in the Schools. In G.H. Sherwood (Ed.), *Free education by the American museum of natural history in public schools and colleges; history and status of museum instruction and its extension to the schools of Greater New York and vicinity* (5–6). New York: Miscellaneous Publications of The American Museum of Natural History, No.13.

Osborne, R.A.L. 1988. *Dreamtime to dust: Australia's fragile environment*. Sydney: Australian Museum.

OUMNH. 2016a. *Dinosaurs in the Museum*. Available at: www.oum.ox.ac.uk/learning/htmls/dinosaur.htm [Accessed 12 January 2016].

OUMNH. 2016b. *Oxfordshire Minerals*. Available at: www.oum.ox.ac.uk/learning/htmls/oxmin.htm [Accessed 12 January 2016].

Owen, R. 1841. *Report on British Fossil Reptiles*. London: Richard and John Taylor.

Owen, R. 1862. *On the extent and aims of a national museum of natural history: including the substances of a discourse on that subject, delivered at the Royal Institution of Great Britain, on the evening of Friday, April 26, 1861*. London: Saunders, Otley, & Co.

Owen, R. 1866. *On the anatomy of vertebrates, Volume 1*. London: Longmans, Green, and Co.

Palmatier, R.A. 1995. *Speaking of Animals: A Dictionary of Animal Metaphors*. Westport, CT: Greenwood Press.

Panjabi, R.K.L. 1997. *The Earth Summit at Rio: Politics, Economics, and the Environment*. Boston, MA: Northeastern University Press.

Parco della Preistoria. 2016. *Home*. Available at: www.parcodellapreistoria.it [Accessed 21 June 2016].

Parham, J. 2016. *Green Media and Popular Culture: An Introduction*. Basingstoke: Palgrave Macmillan.

Parc Préhistorique. 2016. Homepage. Available at: www.prehistoire.com/pages/parc/le_parc.php [Accessed 21 August 2016].

Parr, A.E. 1959. *Mostly about Museums*. New York, NY: American Museum of Natural History.

Patel, K.K. 2012. Integration by Interpellation: The European Capitals of Culture and the Role of Experts in European Union Cultural Policies. *Journal of Common Market Studies,* 51(3): 538–554.

Pattullo, P. 1997. Reclaiming the heritage trail: culture and identity. In L. France (Ed.), *The Earthscan Reader in Sustainable Tourism* (135–147). London: Earthscan.

PC 2197. 1885. Order-in-Council.

Peale, R. 1803. *An historical disquisition on the mammoth: or, great American incognitum, an extinct, immense, carnivorous animal, whose fossil remains have been found in North America*. London: E. Lawrence.

Pearson, K.A. 2007. Beyond the Human Condition: An Introduction to Deleuze's Lecture Course. *Substance,* 36(3): 57–71.

Pech, M-E. 2014. Éducation: le mammouth à la peau dure. *Le Figaro*, 9 May.

Pena dos Reis, R. and Helena Henriques, M. 2009. Approaching an Integrated Qualification and Evaluation System for Geological Heritage. *Geoheritage,* 1(1): 1–10.

Perkinson, H.J. 1991. *Getting Better: Television and Moral Progress*. New Brunswick, NJ: Transaction Publishers.

Perks, S. 2015. Transforming the Natural History Museum in London: Isotype and the New Exhibition Scheme. In *Museum Media. The International Handbooks of Museum Studies, Volume 3* (389–418). Malden, MA: Wiley Blackwell.

Peter Bailey, A. 1975. New York Beat. *Jet*, 24 April, 55.

Petrified Forest of Lesvos. 2016. *Μουσείο*. Available at: www.petrifiedforest.gr/?page_id=959 [Accessed 3 August 2016].

Pick, A. and Narraway, G. (Eds.). 2013. *Screening Nature: Cinema beyond the Human*. New York and Oxford: Berg.

Pinel, P. 1791, *Mémoire lu à la Société d'Histoire Naturelle, sur les moyens de préparer les quadrupèdes, & les oiseaux destinés à former des collections d'histoire naturelle*. Paris: s.n.

PNR du Vexin. 2016. *Sentiers du patrimoine*. Available at: www.pnr-vexin-francais.fr/fr/education-et-culture/valorisation-patrimoines/sentiers-du-patrimoine [Accessed 27 May 2016].

Pocock, D. 1997. Some reflections on World Heritage. *Area,* 29(3): 260–268.

Poliquin, R. 2012. *The Breathless Zoo: Taxidermy and the Cultures of Longing*. Philadelphia: The Pennsylvania State University Press.

Popper, K. 1962. *The Logic of Scientific Discovery*. London: Hutchinson & Co.

Prater, S.H. 1928. Modern Museum Methods. *Journal of the Bombay Natural History Society,* 32: 532–544, 762–772.

Préhisto Dino Parc. 2016. *Les dinosaures*. Available at: www.prehistodino.com/les-dinosaures/ [Accessed 23 June 2016].

Prop., 1909. *125 Kungl. Maj:ts nådiga proposition till Riksdagen angående åtgärder till skyddande af naturminnesmärken å kronans mark samt afsättande af vissa nationalparker*. Stockholm: Government of Sweden.

Quinio, P. 1997. *Allègre, le monsieur muscle du "mammouth". Le ministre de l'Education tente de calmer l'irritation des syndicats*. Available at: www.liberation.fr/france-archive/1997/06/26/allegre-le-monsieur-muscle-du-mammouth-le-ministre-de-l-education-tente-de-calmer-l-irritation-des-s_208696 [Accessed 12 August 2016].

RACCE. 2010. *Home*. Available at: racce.nhmc.uoc.gr [Accessed 3 September 2016].

Rathburn, R. 1913. *A descriptive account of the building recently erected for the departments of natural history of the United States National Museum.* Washington, DC: Government Printing Office.

Red Natura 2000, la vida en los espacios protegidos de España. 2016. La 2 / Natura 2000 [Documentary].

Reichenbach, H.G.L. 1830. *Flora Germanica Excursoria.* Lipsiae: Apud Carolum Cnobloch.

Reichenbach, H.G.L. 1836. *Das Königlich Sächsische Naturhistorische Museum in Dresden. Ein Leitfaden.* Liepzig: Berlag der Magnerschen Buchandlung.

Revkin, A.C. 1992. *Global Warming: Understanding the Forecast.* New York: Abbeville.

Richards, M. 2013. Global Nature, Global Brand: BBC Earth and David Attenborough's Landmark Wildlife Series. *Media International Australia,* 146(1): 143–154.

Ricoeur, P. 2004. *Memory, History, Forgetting.* Chicago: The University of Chicago Press.

Rivière, G.H. 1989. *La Muséologie : selon Georges Henri Rivière: cours de muséologie, textes et témoignages.* Paris: Dunod.

Roberts, T.S. 1922. *Zoological Museum of the University of Minnesota, Annual Report for the year ending June 30, 1922.* Minneapolis, MN: University of Minnesota.

Rogers, E.S. 1969. *Forgotten peoples: a reference.* Ontario: Royal Ontario Museum.

Rössler, P. 2014. *The Bauhaus and Public Relations: Communication in a Permanent State of Crisis.* New York and London: Routledge.

Rough Sea at Dover. 1895. R.W. Paul and B. Acres (Dirs.) [Film].

RTVE. 2016a. *Espacios naturales.* Available at: www.rtve.es/alacarta/videos/espacios-naturales/espacios-naturales-reservas-biosfera-andalucia-marruecos/3530641/ [Accessed 19 August 2016].

RTVE. 2016b. *Red Natura 2000, la vida en los espacios protegidos de España.* Available at: www.rtve.es/television/red-natura-2000/ [Accessed 21 August 2016].

Rumphius, G.E. 1705. *D'Amboinsche rariteitkamer.* Amsterdam: François Halma.

Rüsen, J. 1990. *Zeit und Sinn: Strategien historischen Denkens.* Frankfurt a. M.: Fischer Taschenbuch Verlag.

Said, E. 1978. *Orientalism.* New York: Pantheon.

San Andreas. 2015. B. Peyton (Dir.). Warner Bros. Pictures [Film].

San Francisco. 1936. W. S. Van Dyke and D. W. Griffith (Dirs.). Metro-Goldwyn-Mayer [Film].

Sandell, R. 2016. *Museums, Moralities and Human Rights.* Abingdon and New York: Routledge.

Sandell, R, and Nightingale, E. 2012. *Museums, Equality and Social Justice.* Abingdon and New York: Routledge.

Sandford, L. 1974. *Report of the National Parks Policy Review Committee.* London: DOE, HMSO.

Santa Barbara Museum of Natural History. 2016. *Gladwin Planetarium.* Available at: www.sbnature.org/gladwin/3.html [Accessed 7 June 2016].

Saunders, J.R. 1952. *The world of natural history, as revealed in the American Museum of Natural History.* New York: Sheridan House.

Saurier Park. 2016. *Home.* Available at: www.saurierpark.de [Accessed 13 April 2016].

Sawyer, J.S. 1972. Man-made Carbon Dioxide and the "Greenhouse" Effect. *Nature,* 239: 23–26.

Schama, S. 1995. *Landscape and Memory.* London: Harper Collins.

Scheiner, S.M. and Willig, M.R. (Eds.). 2011. *The Theory of Ecology.* Chicago, IL: The University of Chicago Press.

Schneider, S. 2016. Communicating Uncertainty: a challenge for science communication. In J.L. Drake, Y.Y. Kontar, J.C. Eichelberger, S.T. Rupp and K.M. Taylor (Eds.), *Communicating Climate-Change and Natural Hazard Risk and Cultivating Resilience* (267–278). Dordrecht: Springer.

Schofield, J. and Szymanski, R. (Eds.). 2011. *Local Heritage, Global Context: Cultural Perspectives on Sense of Place*. Farnham: Ashgate.

Schomburgk, R. 1840. *The Guiana exhibition is now open at 209, Regent Street: containing among other curiosities, a coloured drawing, the size of nature*. London: R.H. Schomburgk.

Scotland's Geodiversity Charter. 2012. *Charter.* Available at: https://scottishgeodiversity forum.org/charter/[Accessed 7 September 2016].

Scott, K.D. 2010. The Role of "Spectacle" in Contemporary Wildlife Documentary. *Journal of Popular Film and Television,* 31: 29–35.

Scott, M. 2006. *Rethinking Evolution in the Museum: Envisioning African Origins*. London and New York: Routledge.

Scottish Natural Heritage. 2010. *Siccar Point: Site Management Statement*. Available at: http://gateway.snh.gov.uk/sitelink/documentview.jsp;jsessionid=50b42adc823c375 01edfb96058d4b6cc53ef0f7440ac6fee69ecdade3baa2b93.e38KahaMax4Rai0Oax8 Sb3mLc350?p_pa_code=1432&p_Doc_Type_ID=3 [Accessed 19 July 2016].

Scudder, J. 1823. *A Companion to the American Museum; Being a Catalogue of Upwards of Fifty Thousand Natural and Foreign Curiosities, Antiquities, and Productions of the Fine Arts: Now Open for Public Inspection, in the New-York Institution, Park, Broadway; with the Latin Name of Each Natural Curiosity Prefixed. To which is Added, Notes and Explanatory Remarks*. New York: G.F. Hopkins.

Sedgewick, A. 1821. *A Syllabus of a Course of Lectures on Geology*. Cambridge, England: J. Hodson.

Sellars, R.W. 1997. *Preserving Nature in the National Parks: A History*. New Haven, CT: Yale University Press.

Sendero Señalizado Los Yesares. 2016. *Paraje Natural Karst en Yesos de Sorbas*. Available at: www.juntadeandalucia.es/medioambiente/servtc5/ventana/mostrarFicha.do?re=s&id Equipamiento=19362 [Accessed 17 August 2016].

Semonin, P. 2000. *American Monster: How the Nation's First Prehistoric Creature Became a Symbol of National Identity*. New York and London: New York University Press.

Senckenberg Naturmuseum. 2016. *Evolution des menschen*. Available at: www.senckenberg. de/root/index.php?page_id=2556 [Accessed 17 August 2016].

Sheets-Johnstone, M. 1999. *The Primacy of Movement*. Philadelphia: John Benjamins.

Sheets-Pyenson, S. 1988a. How to grow a natural history museum: the building of colonial collections, 1850–1900. *Archives of Natural History,* 15(2): 121–147.

Sheets-Pyenson, S. 1988b. *Cathedrals of Science: The Development of Colonial Natural History Museums during the late 19th century*. Montreal: McGill-Queen's University Press.

Sherwood, G.H. 1920. *Free education by the American museum of natural history in public schools and colleges; history and status of museum instruction and its extension to the schools of Greater New York and vicinity*. New York: Miscellaneous Publications of The American Museum of Natural History, No.13,

Simba: The King of the Beasts. 1928. M. Johnson and O. Johnson (Dirs.). Martin Johnson African Expedition Corporation (USA) [Film].

Singh, A. 2015. Save Dippy: outcry over Natural History Museum plan to eject famous dinosaur. *The Daily Telegraph*, January 29.

Smith, A.R. 1848. *The natural history of the idler upon town*. London: Vicetelly Brothers.

Smith, L. 2006. *Uses of Heritage*. London and New York: Routledge.

Smith, L. and Campbell, G. 2015. The elephant in the room: heritage, affect and emotion. In W. Logan, M. Nic Craith and U. Kockel (Eds.), *A Companion to Heritage Studies* (443–460). New York: Wiley.

Smith, N. 1984. *Uneven Development: Nature, Capital and the Production of Space*. Oxford: Blackwell.

SNMNH. 2016a. *Remake: Our iconic elephant has been reimagined*. Available at: http://naturalhistory.si.edu/exhibits/elephant/ [Accessed 19 March 2016].

SNMNH. 2016b. *The Evolving Universe*. Available at: http://naturalhistory.si.edu/exhibits/evolving-universe/science/ [Accessed 19 March 2016].

Société Géologique de France. 1991. *Déclaration internationale des droits de la mémoire de la Terre*. Available at: www.geosoc.fr/sommaires-et-resumes/cat_view/77-patrimoine-geologique.html [Accessed 20 May 2016].

Société Géologique de France. 1994. *Proceedings of the 1st International Symposium on the Conservation of our Geological Heritage (Digne-les-Bains, 11–16 June 1991)*. Mémoires de la Soc. Géol. de France, Nouvelle Série, No. 165.

Solli, B., Burström, M., Domanska, E., Edgeworth, M., González-Ruibal, A., Holtorf, . . . and Witmore, C. 2011. Some Reflections on Heritage and Archaeology in the Anthropocene. *Norwegian Archaeological Review*, 44(1): 40–88.

Soulé, M.E. and Lease, G. (Eds.). 1995. *Reinventing Nature? Responses to Postmodern Deconstruction*. Washington, DC: Island Press.

South Downs National Park. 2016. *Geology of the South Downs*. Available at: www.southdowns.gov.uk/discover/landscape-geology/geology-of-the-south-downs/ [Accessed 7 September 2016].

Spary, E.C. 2000. *Utopia's Garden: French Natural History from Old Regime to Revolution*. Chicago, IL and London: University of Chicago Press.

Speakman, C. 2001. *Yorkshire Dales: The Official National Park Guide*. Bath: Crimson Publishing.

Spencer, H. 1864. *The Principles of Biology*. London: William and Norgate.

State of the Planet. 2000. BBC Productions [Documentary].

Statens Naturhistoriske Museum. 2016. *Museum History*. Available at: http://snm.ku.dk/omsnm/museetshistorie/ [Accessed 7 September 2016].

Steinberg, M.P. 1996. Cultural History and Cultural Studies. In C. Nelson and U. Parameshwar Gaonkar (Eds.). *Disciplinarity and Dissent in Cultural Studies* (103–129). New York and London: Routledge.

Storey, D. 2010. Using the past: heritage and re-imagining rural places. In D.G. Winchell, D. Ramsey, R. Koster and G.M. Robinson (Eds.), *Geographical Perspectives on Sustainable Rural Change* (374–383). Brandon, MB: Rural Development Institute, Brandon University.

Strahan, R. 1979. *Rare and Curious Specimens: An Illustrated History of the Australian Museum*. Sydney: Australian Museum.

Swainson, W. 1840. *Taxidermy, bibliography and biography*. London: Longman, Brown, Green, Longmans, & Roberts.

Sykes, G. 1977. Exhibition: Mankind. The Natural History Museum. *Spare Rib*, 63: 45.

Tarzan of the Apes. 1918. S. Sidney (Dir.). Hollywood Film Enterprises [Film].

Taylor, D.E. 2016. *The Rise of the American Conservation Movement: Power, Privilege, and Environmental Protection*. Durham, NC: Duke University Press

Taylor, J.E. 1885. *Our Common British Fossils and where to Find Them: A Handbook for Students*. London: Chatto and Windus.

Thal, C. 1921. *The Early History of the Museum. Year Book of the Public Museum of the City of Milwaukee*. Milwaukee: Public Museum of the City of Milwaukee, Board of Trustees.

Thamdrup, H.M. 1947. The Mols Laboratory. An Ecological Laboratory and its working Programme. *Natura Jutlandica* 1: 67–133.

Thamdrup, H.M. 1978. Naturhistorisk Museum med Molslaboratoriet. *Acta Jutlandica* 51: 499–519.

The Devil at 4 O'clock, 1961. M. LeRoy (Dir.). Columbia Pictures [Film].

The Giant's Causeway. 2016. *Home*. Available at: www.nationaltrust.org.uk/giants-causeway [Accessed 7 September 2016].

The Hurricane, 1937. J. Ford (Dir.). United Artists [Film].

The Land of the Living Dinosaurs. 2016. West Midlands Safari Park. Available at: www.wmsp.co.uk/dinosaurs/ [Accessed 20 September 2016].

The Life of Birds, 1998. BBC Productions [Documentary].

The Life of Mammals, 2002. BBC Productions [Documentary].

The Living Planet, 1984. BBC Productions [Documentary].

The Lost World, 1925. H. Hoyt (Dir.). First National Pictures [Film].

The Private Life of Gannets, 1934. J. Huxley (Dir.). London Film Productions [Film].

The Rains Came, 1939. C. Brown (Dir). 20th Century Fox [Film].

The Trials of Life, 1990. BBC Productions [Documentary].

The National Trust for Scotland, 2016. Dual World Heritage Status For Unique Scottish Islands. Available at: www.kilda.org.uk/frame26.htm [Accessed 22 June 2016].

The Natural History Museum, 2016. About us. Available at: http://thenaturalhistorymuseum.org/about/ [Accessed 13 May 2016].

The Natural Trust for Scotland, 2016. St Kilda. Available at: www.kilda.org.uk/ [Accessed 22 June 2016].

The Private Life of Gannets, 1934. J. Huxley (Dir.). London Film Productions [Film].

The Rains Came, 1939. C. Brown (Dir). 20th Century Fox [Film].

The Trials of Life, 1990. BBC Productions [Documentary].

The Triumph of Man. 1964. *New York World's Fair*. New York: The Travelers Insurance Companies / RCA Records.

The Wilderness Act. 1964. *Public Law 88–577 (16 U.S.C. 1131–1136. 88th Congress, Second Session. September 3, 1964*. Washington, DC: Government Printing Office.

Thoreau, H.D. 1854. *Walden; or, Life in the woods*. Boston: Ticknor and Fields.

Three Ages, 1923. B. Keaton and E.F. Cline (Dirs.). Metro Pictures [Film].

Thrift, N. 2000. Afterwords. *Environment and Planning D: Society and Space, 18*: 213–255.

Thrift, N. 2003. Performance and. . . . *Environment and Planning A*, 35: 2019–2024.

Todmorden Moor, 2013. Geology and Heritage Trail. Available at: www.todmordenmoor.org.uk/trail.html [Accessed 20 May 2016].

Toscan, G. 1795. *Histoire du Lion de la Ménagerie du Museum national d'histoire naturelle*. Paris: Museum national d'histoire naturelle.

Tradescant, J. 1656. *Musaeum Tradescantianum, or, A collection of rarities preserved at South-Lambeth neer London*. London: John Grismond.

Tuan, Y. 1977. *Space and Place: The Perspective of Experience*. Minneapolis: University of Minnesota Press.

Tudge, C. 1991. *Global Ecology*. London: Natural History Museum.

Tully, C. 2002. *The Broads: The Official National Park Guide*. Bath: Crimson Publishing.

UCMP, 2015. The *T-Rex* Expo: Building the Perfect Beast. Available at: www.ucmp. berkeley.edu/trex/ [Accessed 13 May 2016].

UN. 1992. *United Nations Framework Convention on Climate Change*, FCCC/ INFORMAL/84. GE.05–62220 (E). 200705. Signed and authorised 9 May 1992. Available at: https://unfccc.int/resource/docs/convkp/conveng.pdf [Accessed 12 September 2016].

UN. 1997. *Kyoto Protocol*. Available at: http://unfccc.int/kyoto_protocol/items/2830.php [Accessed 10 March 2016].

UN. 2015. *Paris Agreement*. Available at: http://unfccc.int/paris_agreement/items/9485. php [Accessed 10 March 2016].

UNCED. 1992. *Rio Declaration on Environment and Development*. Available at: www.unep. org/documents.multilingual/default.asp?documentid=78&articleid=1163 [Accessed 16 July 2016].

UNEP. 1972. *Report of the United Nations Conference on the Human Environment*. Available at: www.unep.org/Documents.Multilingual/Default.asp?DocumentID=97 [Accessed 19 July 2016].

UNESCO. 1972. *Convention Concerning the Protection of the World Cultural and Natural Heritage*. Available at: whc.unesco.org/en/conventiontext [Accessed 22 July 2016].

United States Congress. 1872. *An Act to set apart a certain tract of land lying near the headwaters of the Yellowstone River as a public park, 42nd Congress, 2nd Session, 1871, c. 24, 17 STAT., 32 and 33*. Washington, DC: Government Printing Office.

United States National Museum. 1919. *Natural history building*. Washington, DC: Government Printing Office.

United States National Museum. 1959. *Annual Report for the Year Ended June 30, 1959*. Washington, DC: Smithsonian Institution.

Urban Science: The Birth of a Flower. 1910. P. Smith [Photographer]. Kineto [Film].

Van Praët, M. 2012. *Reinventing the Museum to Mankind: Museums in an Age of Migrations*. Milan: Mela Books.

Veyrat-Masson, I. 2000. *Quand la télévision explore le temps. L'histoire au petit écran. 1953–2000*. Paris: Fayard.

Viveiros de Castro, E. 2014. *Cannibal Metaphysics*. Minneapolis, MN: University of Minnesota Press.

VMNH. 2016. *The Harvest Foundation Hall of Ancient Life*. Available at: www.vmnh.net/ the-harvest-foundation-of-the-piedmont-great-hall [Accessed 17 May 2016].

Vogel, S. 1996. *Against Nature: The Concept of Nature in Critical Theory*. Albany, NY: SUNY Press.

Voghera, A. 2011. *After the European landscape convention: policies, plans and evaluation*. Florence: Alinea.

Vogt, V.L. 2009. Wandern und Trekking als Freizeitaktivität und Marktsegment im Naturtourismus. *Naturschutz und Landschaftsplanung*, 41(8): 229–236.

Voyage au bout du monde. 1976. P. Cousteau (Dir.). Cousteau Group [Film].

Vuillemin, N. 2009. *Les beautés de la nature à l'épreuve de l'analyse: programmes scientifiques et tentations esthétiques dans l'histoire naturelle du XVIIIe siècle, 1744–1805*. Paris: Presses Sorbonne Nouvelle.

Walking with Beasts. 2001. BBC Productions [Documentary].

Walking with dinosaurs. 1999. BBC Productions [Documentary].

Wallace, M. 1996. *Mickey Mouse History and Other Essays on American Memory*. Philadelphia: Temple University Press.

Ward, H.A. 1870a. *Catalogue of the academy series of casts of fossils: from the principal museums of Europe and America, with short descriptions and illustrations.* Rochester, NY: E. R. Andrews.

Ward, H.A. 1870b. *Catalogue of the College Series of Casts of Fossils: From the Principal Museums of Europe and America, with short descriptions and illustrations.* Rochester, NY: E. R. Andrews.

Wastl, J. and Sittenberger, A. 1941. Rassenkundliche Untersuchungen an Deutschen und Tschechen im südlichsten Böhmerwald (Quellgebiet der Moldau). *Annalen des Naturhistorischen Museums in Wien,* 52: 397–457.

Waters, C.N., Zalasiewicz, J., Summerhayes, C., Barnosky, A.D., Poirier, C., Gałuszka, . . . and Wolfe, A.P. 2016. The Anthropocene is functionally and stratigraphically distinct from the Holocene. *Science,* 351: 6269.

Weale, A. 2005. Environmental Rules and Rule-making in the European Union. In A. Jordan (Ed.), *Environmental Policy in the European Union: Actors, Institutions, and Processes* (125–140). Sterling, VA: Earthscan.

Weik von Mossner, A. 2011. Reframing Katrina: The Color of Disaster in Spike Lee's When the Levees Broke. *Environmental Communication,* 5(2): 146–165.

Wellock, T.R. 2007. *Preserving the Nation: The Conservation and Environmental Movements 1870–2000.* New York: J. Wiley.

Wells, W. 1960. Our technological dilemma or an appraisal of man as a species bent on self-destruction. *Bulletin of the Atomic Scientists,* 16(9): 362–365.

Wertsch, J. 2002. *Voices of Collective Remembering.* Cambridge, England: Cambridge University Press.

West, S. (Ed.). 2010. *Understanding Heritage in Practice.* Manchester: Manchester University Press.

Westphal, B. 2007. *La Géocritique. Réel, fiction, espace.* Paris : Éditions.

Wheatley, H. 2004. The Limits of Television? Natural History Programming and the Transformation of Public Service Broadcasting. *European Journal of Cultural Studies,* 7(3): 325–339.

Whewell, W. 1837. *History of the Inductive Sciences: from the earliest to the present times, Vol.1.* London: John W. Parker.

White, A.M. 1953. Eighty-Fifth Annual Report of the President. In American Museum Of Natural History (Ed.), *The American Museum of Natural History, Eighty-Fifth Annual Report* (3–7). New York, NY: American Museum of Natural History.

White, H. 1973. *Metahistory: The Historical Imagination in Nineteenth-Century Europe.* Baltimore, MD: Johns Hopkins University Press.

White, K. 1994. *Le Plateau de l'Albatros: Introduction à la géopoétique.* Paris: Grasset.

White, K. 1996. *Le Livre des abîmes et des hauteurs.* Paris : Editions.

Whitlock, H.P. 1918. Minerals and Gems. In American Museum Of Natural History (Ed.), *Fiftieth Annual Report* (61–63). New York: American Museum Of Natural History.

Wiber, M. 1998. *Erect Men/Undulating Women: The Visual Imagery of Gender, Race and Progress in Reconstructive Illustrations of Human Evolution.* Waterloo, Ontario: Wilfrid Laurier University Press.

Wild China. 2008. BBC Productions [Documentary].

Wilson, A. 1991. *The Culture of Nature: North American Landscape from Disney to the Exxon Valdez.* Toronto: Between the Lines.

Wilson, R. 2014. Playful Heritage: excavating Ancient Greece in New York City. *International Journal of Heritage Studies,* 21(5): 476–492.

Wilson, R. 2016. *The Language of the Past.* London: Bloomsbury.

Winter, D. 1996. *Ecological psychology: Healing the Split between Planet and Self.* New York: Harper Collins.

Winter, J.M. 1992. *Sites of memory, sites of mourning.* Cambridge, England: Cambridge University Press.

Winter, T. 2013. Clarifying the critical in critical heritage studies. *International Journal of Heritage Studies,* 19(6): 532–545.

Wonders, K. 1993. *Habitat Dioramas: Illusions of Wilderness in Museums of Natural History.* Uppsala: Acta Universitatis Upsaliensis.

World Commission on Environment and Development (WCED). 1987. *Our Common Future.* Oxford: Oxford University Press.

World Museum. 2016. *Planetarium.* Available at: www.liverpoolmuseums.org.uk/wml/events/planetarium-listings.aspx [Accessed 19 July 2016].

Worster, D. 1994. *Nature's Economy: A History of Ecological Ideas.* Cambridge, England: Cambridge University Press,

Wright, P.M. 1985. *On Living in an Old Country: The National Past in Contemporary Britain.* London: Verso.

Wurzel, R.K.W. 2002. *Environmental Policy-Making In Britain, Germany and the European Union.* Manchester: Manchester University Press.

Wurzel, R.K.W. 2004. *Germany: from environmental leadership to partial mismatch.* In A. Jordan and D. Liefferink (eds) *Environmental Policy in Europe* (99–117). London and New York: Routledge.

Yale Peabody Museum of Natural History. 2016a. *Mission & History.* Available at: http://peabody.yale.edu/about-us/mission-history [Accessed 19 May 2016].

Yale Peabody Museum of Natural History. 2016b. *The Age of Mammals Mural.* Available at: http://peabody.yale.edu/exhibits/age-mammals-mural [Accessed 19 May 2016].

Yale Peabody Museum of Natural History. 2016c. *Fossil Fragments: the riddle of human origins.* Available at: http://peabody.yale.edu/exhibits/fossil-fragments/home [Accessed 19 May 2016].

Yale Peabody Museum of Natural History. 2016d. *The Hall of Minerals, Earth and Space.* Available at: http://peabody.yale.edu/exhibits/hall-minerals-earth-and-space [Accessed 19 May 2016].

Yanni, C. 1999. *Nature's Museums: Victorian Science and the Architecture of Display.* New York: Princeton Architectural Press.

Yapp, G.W. and Ellis, R.F.L.S. (Eds.). 1851. *Official Catalogue of the Great Exhibition.* London: Spicer Brothers.

ZDF. 2016a. *Terra X.* Available at: www.zdf.de/dokumentation/terra-x [Accessed 3 August 2016].

ZDF. 2016b. *Geheimnisse der Tiefsee.* ZDF. [Documentary].

ZDF. 2016c. *The Desert Sea.* ZDF [Documentary].

ZDF. 2016d. *The Amazon of the East.* ZDF [Documentary].

Ziegler, W. 1988. *Natural History Museum Senckenberg guide.* Frankfurt: Senckenberg Nature Research Society.

Zimmer, C. 1931. The Natural History Museum (Das Museum fur Naturkunde), Berlin. *Natural History Magazine,* 4: 353.

Žižek, S. 2011. *Living in the End Times.* London: Verso.

Žižek, S. 2014. *Trouble in Paradise: From the End of History to the End of Capitalism.* London: Allen Lane.

Żylicz, T. 1994. Implementing Environmental Policies in Central and Eastern Europe. In A.M. Jansson, M. Hammer, C. Folke and R. Costanza (Eds.), *Investing in Natural Capital: The Ecological Economics Approach to Sustainability* (408–430). Washington, DC: Island Press.

Index

Milton Keynes UK
Ingram Content Group UK Ltd.
UKHW040101071024
449327UK00019B/729

9 780367 244125